Good in a Bed

Other Books by the author

An Anthology of Garden Writing
The Pleasures of Gardening
The Classic Horticulturist (with Nigel Colborn)
Foliage Plants
The Village Show
Wall Plants and Climbers
The Garden Book (with David Stevens)
Gardening for Pleasure
Plants for All Seasons

Good in a Bed

Garden Writings from *The Spectator*

URSULA BUCHAN

JOHN MURRAY
Albemarle Street, London

Introduction, selection and editorial matter © Ursula Buchan 2001
Articles © *The Spectator* 1984, 1985, 1986, 1987
Articles © Ursula Buchan 1987, 1988, 1989, 1990, 1991, 1992, 1993,
1994, 1995, 1996, 1997, 1998, 1999, 2000

Illustrations © Timothy Jaques 2001

First published 2001
by John Murray (Publishers) Ltd,
50 Albemarle Street, London W1S 4BD

A catalogue record for this book is available from the British
Library

ISBN 0-7195-6026 8

Typeset in Adobe Palatino 11/14 pt by
Servis Filmsetting Ltd, Manchester
Printed and bound in Great Britain by
St Edmundsbury Press Ltd,
Bury St Edmunds, Suffolk

*This book is affectionately dedicated to my brothers and sisters –
Deborah Stewartby, Toby Buchan, Edward Buchan, Laura Chanter
and James Buchan – whose loyalty as readers is only equalled by
their forebearance as critics.*

Contents

Contents

Acknowledgements

My association with *The Spectator* since 1984 has been so much fun that it is good to have the chance to say how lucky I have been in both my editors – Charles Moore, Dominic Lawson, Frank Johnson and Boris Johnson – and Arts section editors – Gina Lewis, Jenny Naipaul, Rebecca Nicolson and Elizabeth Anderson. Their tolerance of my last-minute, and minute, changes has been nothing short of heroic; but, best of all, they have allowed me to write what I wanted, and scarcely ever changed or cut anything, which makes them well-nigh unique among editors on national newspapers and magazines. This attitude is no doubt encouraged by the fact that *The Spectator* carries few pictures, so layout can accommodate the whims of writers, but I believe it has more to do with a flattering and self-fulfilling assumption that, if I wrote something, it was because I meant to.

I should also like to acknowledge those readers of *The Spectator* who have taken the trouble to write to me over the years and who are always courteous, well-informed and appreciative of a joke, however corny. I particularly treasure the letter I received from a Spaniard who told me that he had lived in this country for some years and read *The Spectator* partly in order to improve his English. What, he enquired, did I mean, when writing about a public committee going slow on a project, by the expression 'dropping a dead bat'?

Acknowledgements

As far as this book is concerned, I must pay tribute to the sterling efforts on my behalf of Michael Shaw and Jonathan Pegg at Curtis Brown. It has been a great pleasure to work with the team at John Murray, in particular Caroline Knox, Gail Pirkis, Liz Robinson and Hilary Bird. I am immensely grateful to Tim Jaques, who has illustrated the book so wittily, and thoroughly entered into the spirit of the thing. I have completely forgiven him for the picture on the jacket, although I am not sure I can say the same for Kipling, my yellow Labrador. Last, but never least, I must thank my husband, Charlie, and our children, Emmy and Tom, for their interest and encouragement. They all have the great merit for me of not being keen gardeners. By writing to please them, I hope thereby to please other readers as well.

Introduction

I always try to write for a readership with a variety of interests, of which gardening may or may not be one. I want most to communicate something of the deep attachment I have for the subject, and the pleasure, both practical and intellectual, which I take in it. I have a number of preoccupations, some normal, some barmy, and assume that the reader has too. Gardening, and the people who do it, are part of the rich texture of life, and therein lies much of the fun of it. Seeing the subject only in its own terms can make for jolly dull reading, as anyone who buys gardening magazines may well have discovered. I am always conscious of the fact that the article I am writing could be the first that someone has ever read on gardens and gardening and, if I make a hash of it, it may well be their last.

I stumbled into garden writing about twenty years ago, mainly because I didn't have a deep enough voice to sound convincing as anyone's Head Gardener, after five years' practical and theoretical training in horticulture: in 'private service' (see page 85); at the Royal Horticultural Society's Gardens at Wisley; in a bulb firm in Holland and an arboretum in Belgium; and, above all, at Kew. While I was at Kew I began writing the occasional freelance article in journals and magazines, and in 1980, after moving with my husband to Northamptonshire, became for a short while the gardening correspondent of the *Peterborough Evening Telegraph*. So began what has been a hugely enjoyable career, never entirely interrupted by the arrival of two children in the early 1980s and a change of house (to acquire a big garden) in 1993 (see page 18).

In 1984 I wrote to Charles Moore, the newly-installed editor of *The Spectator*. I suggested he might consider introducing a gardening column, even though I was conscious that this political and literary weekly magazine had done perfectly well without one since its foundation in 1828. By chance, Charles had lately seen market research which revealed that only a minority of the magazine's readers were women. Not unnaturally, since it is true, he assumed that women, as well as men, enjoyed reading about gardening. So he took a gamble on an almost untried and totally unknown writer, and began an association which has lasted seventeen years, and spanned the tenure of four editors. So far.

This book is a collection of some of the articles which have appeared since 1984 under the title 'Gardens' in *The Spectator*. The purpose is to introduce them to those who have not seen them, and to reacquaint *Spectator* readers with some that they may have enjoyed first time round but which have been irretrievable since. The pieces have been selected as much on their literary as their horticultural merits, and to provide a breadth of subject-matter. I have concentrated on those which I feel have best stood the test of time; the relevance of some has of course disappeared entirely in the rush and press of subsequent events, and these have been disregarded except in one or two instances when a period piece seemed worth reproducing. I have also tended to avoid what one might call 'political' articles, those concerned with Lottery projects, for example, or the Royal Horticultural Society's battle over the siting of its Lindley Library; although possibly of some historical interest one day, they look like pretty stale buns right now. That said, it is surprising how many of the articles have not dated badly, and are even still perfectly topical. That is partly due to the eternal verities of the subject, and partly because trends tend to be long-term. Where I have thought it interesting to continue the story, or to explain something which over time has lost its context, I have done so. In places I have also edited the articles, mainly because some expressions have become old-fashioned even in the comparatively short time I have been writing. I have, for example, changed 'he' to 'she' when talking about 'the gardener', because women now make up the majority of gardeners in this country. All that said, the articles are essentially as they first appeared.

I am only sorry that it is well-nigh inevitable that the pages should be peppered with Botanical Latin names, which act as nasty visual doorstops, liable to trip up the reader. Gardening terms can also militate against an easy read, for ordinary words become terms of art, with different and often very specific meanings. In order to avoid distracting lunacies like '20m/66ft', however, conversion charts have been provided on page 251. As for any plant name change which has occurred in the intervening time since the article was written, unless there is likely to be misunderstanding I have been reluctant to alter the text. After all, by the second edition of this book they may have changed back again!

Gardening, both public and private, has changed enormously in the last twenty years and I hope that I have changed with it. For example, there is far more emphasis on the conservation of historic gardens, and their accessibility to everyone. Lottery largesse has altered perspectives and prospects. Television programmes have transformed the way garden owners treat the space they look after. The range of plants readily available to gardeners, thanks to both specialist nurseries and garden centres, has expanded exponentially. Organic gardening has come out of the joke cupboard and is not far short of being the new orthodoxy. Design is going forward again at last, and designers are making gardens which at least have some relevance to the circumstances of those to whom they belong. Garden designers can finally make a living out of a population which better understands their worth. Advances in publishing technology have resulted in new, well-laid-out magazines and sumptuously-illustrated books, which in themselves have fostered the art of garden photography. Only the lack of gardening space and competent, affordable help restrain private gardeners from achieving all their dreams. We are, at the very least, in a silver age. And I, for one, am enjoying it.

1

Me and My Garden

In darker moments, I find it difficult to imagine why any busy person should want to take up gardening. Gardening proper, that is, rather than outdoor housework. To do it really well requires skill, experience, sensitivity, perseverance, reasonable physical fitness, luck and, above all, time. Anyone who wants to garden has to learn to snatch some pretty unforgiving minutes. What's more, most of us begin too late.

In our youth, gardening seems an unwelcome distraction, even irrelevant, yet that is when we should begin. (As it happens, I did, but compelled by circumstances and most reluctantly.) In middle age we will have acquired the garden in which to develop an interest, but by then it may be too late to plant a mulberry tree which will fruit in our lifetime. By the time we have acquired enough knowledge to make a fine garden, and the forces of unrule ranged against us have been tamed, there may not be many years left before the hand on the trowel seizes up, and bending for that weed becomes impossible. The battle for the garden begins once more. In old age, a fleeting scent reminds us that we may never again wander out on a summer's evening to dead-head the roses.

Despite this, we take it up in middle life; and with it a number of anxieties which we could probably do without. Now that the garden is so fashionable as exterior decoration, and everyone has at least a little knowledge, our friends ask difficult questions about colour schemes and plant succession. Fear of embarrassment restricts our entertaining to the winter months, when no one is tempted to ask to see round the garden. Gardening makes us unfit for society anyway: not only can we bring a dinner party to a grinding halt with a sprinkling of inoffensive Latin names, but where two or more keen gardeners are gathered together, silly rivalries often undermine the pleasure of what should be consoling expert chat.

Moreover, at a time when most of us are at full stretch financially, we discover how expensive gardening can be. It is subject to the same modern imperatives and vagaries as fashions in clothes. No longer content with concrete paths and besoms, 'January King' cabbages and our neighbours' cast-off gladioli, we must have York stone paving and garden vacuum cleaners, Italian radicchio and rare fritillaries. We spend fortunes on plants from garden centres, for we have no time to grow anything from seed or cuttings.

Yet, even at my most pessimistic, I cannot ignore the evidence of my eyes: that the busiest people, already encumbered with a variety of interests and preoccupations, will clamp on shapeless hats and coats on Saturday afternoons in winter (when they could have their feet up in front of the Six Nations) and stomp outside to prune the clematis or dig over the vegetable garden. Elegant women, coiffed and manicured, plunge their hands without hesitation into half-rotted kitchen waste. Overworked clergymen spend precious Mondays climbing tall ladders to remove the wisteria from the gutters. Frail but dignified old ladies tiptoe out in terrifying gales to divide their snowdrops. Hard-pressed mothers swap cuttings at the school gate. It is remarkable how many rational people take to gardening and then, despite all reason, refuse to give it up. Just think how often it is named as a recreation in *Who's Who*.

There must be something in it. It would be easy enough to say that it is the lure of lovely flowers or the pleasure of seeing a successful artistic effect. These are certainly very important, but we can appreciate them well enough in other people's gardens, without getting dirt under our fingernails.

In the end, it must be atavism. Just as some deep if unacknowledged instinct will turn a sophisticated weekday boulevardier into a madcap follower of foxhounds on Saturday, so fraught captains and lieutenants of industry, media celebrities, politicians even, cannot wait until the weekend to become medieval villeins, scratching in the scented earth, making something fugitive but enduringly satisfying out of nothing. Perhaps, as Edward Thomas wrote in 'Digging':

> It is enough
> To smell, to crumble the dark earth,
> While the robin sings over again
> Sad songs of Autumn mirth.

I think so.

The following was written at a time when I was doing research into the life and works of Reginald Farrer, among others, for my first book, An Anthology of Garden Writing, *which came out in 1986. The style I wrote this in, almost out-Farrering Farrer, was quite deliberate. At least I think it was.*

Peaks and Troughs   *30 November 1985*

Such is the perversity of human nature that most people, if set down to garden on a heathy moor, will conceive a passionate and unquenchable desire to cultivate roses, carnations and bearded irises. The same people, and I count myself among them, finding themselves living on a belt of limestone, cannot resist the temptation to coax a camellia or pieris to a sickly, unthrifty existence.

That is why, when I was offered a stone trough, I could not contain my excitement at the prospect of isolating some lime-hating alpines from the baleful influence of my alkaline soil. My pleasure was two-fold, for as every grower of alpines knows, the genuine stone trough is now as rare and sought-after as Centre Court tickets on Wimbledon Finals day.

The fault, if you can call it that, lies with the missionary zeal of Clarence Elliott, a nurseryman and plant-hunter who, in the 1930s, popularised the idea of the trough or sink as the proper receptacle for many difficult plants that would curl up at the mention of an ordinary garden soil. In these containers conditions could be so closely controlled, and the plants' capricious wants so admirably cared for, that success was almost assured. The result was that all over the country, sharp-eyed and covetous alpinists hauled troughs from fields under the gaze of bemused farmers, ransacked dumps and rummaged in builders' yards. The supply, never limitless, ran out, and gardeners have since been forced to make their own substitutes, using a mixture of sand, peat and cement stuck onto more modern glazed sinks or even poured into cardboard box moulds. The result, named 'hypertufa', bears as much resemblance to tufa rock as plastic does to leather.

My father's continual efforts on my behalf, to bring the conversation in the pub around to the *recherché* subject of finding and procuring a stone sink, surprisingly met with success. So it was that I found myself in a field in Oxfordshire, sizing up a large and handsome Hornton stone trough. What did I care that an archaeologist of the future might be rather puzzled to find a chunk of dark iron-tinted liassic limestone in an area of creamy-yellow oolite?

The farmer and his son, indulgently prepared to accept £25 for this, to them, worthless lump of stone, took their good nature further and offered to help us load it into a borrowed Land Rover, which act of kindness, when they later counted their strained muscles, they must rather have regretted. For the trough measured five feet long and more than two feet wide, and was very, very heavy.

I enlisted the help of a local handyman to install it. He arrived, unfortunately, while we were away on holiday, so that I was prevented from overseeing its placement in a semi-shaded position below the study window, a deprivation which I shall always regret; for not only would I have enjoyed the prospect of watching the trough sliding down the garden on rollers, aided by Mr S., his sisters and his cousins and his aunts, but I might perhaps have had some influence on its positioning. He built two stone plinths on which to rest it but, with the instinct of adventurous asymmetry which is the hallmark of the bodger, he did not align it equally. My husband, himself no mean bodger, in a fit of neurotic anxiety that his children might pull half a ton of stone down upon their toes, had ordered that the trough be cemented to its supports. It cannot now be moved. I am training myself not to notice, every time I look at this splendid object.

It was the work of several moments to mix together an acid loam, begged from a friend, with large quantities of moss peat and quartzite grit, and a generous helping of bone-meal. The smell of peat, it must be admitted, hung in the air for several days, exhaling a breath of dear old Ireland into the study. The planting was sheer delight: primulas, alpine calceolarias, cassiopes and alpenroses, such as I thought never to see in my garden. These have grown away with a determination and a flourish that leads me to suppose that they are fully patriated.

It is not perhaps always properly appreciated that many alpine plants do not cling with the rugged tenacity of an Everest expedition to sheer cliffs and ledges but are bog plants, for which having their roots in a peaty moisture-retentive mixture is as much as they could wish. This is partly why so many alpines can be persuaded to flourish in humid, lowland gardens.

I am hoping that the iron content of the stone will, at least to some extent, neutralise the alkalinity of the limestone. The growing medium, in any event, is as peaty as an Islay malt whisky. If the leaves of *Rhododendron impeditum* turn the colour of a birch in autumn, I shall be forced to line the trough with zinc, but that is a course of action requiring such dexterity as to be contemplated only in a matter of horticultural life or death.

In case I should be thought boastful (and, after all, tomorrow may bring the untimely death of all the occupants and with them my best hopes), I must say that I regard the trough as a monumental piece of good fortune. There are people far more expert than me who would do anything, stopping only just short of criminality, in order to possess such a thing. For an unworldly creature such as myself, it represents the very peak of earthly ambition.

Doing the Flowers   *12 December 1986*

Christmas is the season of anxiety. In our house the vexed question of whether we think Father Christmas physically capable of carrying a bicycle down the chimney fades into insignificance in comparison with what I am to do about the church flowers.

For me, this issue resolves itself into two principal and almost intractable problems. The first is how to secure my vase safely to the very narrow, sloping sill below the side window which has been allotted to me. Some people worry about aeroplanes dropping out of the sky onto their heads; I worry about my vase coming loose from its moorings and toppling forward during Evensong, drenching Mrs S., who always sits next to the window. By coincidence she wears a hat the shape of an empty flower pot, so there is a sense in which such a disaster might bring about some improvement.

That catastrophe may never happen, but even worse already has. Unlike the east window beyond the altar, the four large side windows are only decorated for the great festivals of the Christian year. One May, confused by a bank holiday which had diverged from Whitsun by the matter of a fortnight, I failed to remember to decorate 'my' window. As I entered the church for the Sunday service, the blood almost froze in my veins. It was the kind of moment which, in old thriller films, received a crashing chord. Instead of an explosion of white lilac and red peony, my window was as empty, as cold and as dreary as Fulham Broadway underground station at midnight. Lacking the nerve to run away, I endured my shame for an hour's eternity. I strongly suspect that village history is now dated by that incident: 'Wasn't that the Whitsun Mrs —— forgot to do her window?'

No less problematical than making a vase safe is what to put in it in this dead, dormant season. When I came to this garden I removed many of the unexciting shrubs and herbaceous perennials which inhabited it, on the ground that *everybody* had them. *I* would grow more unusual, and therefore interesting, plants. So I wantonly destroyed forsythia and pyrethrum, chrysanthemums and Chinese lanterns, only to find I had to replace them (with 'interesting' cultivars, of course) in order to find material for the church at different times of the year. I now see that they may well have bored my predecessor as well, but they were there for a purpose. Most stupidly of all, I dug up a holly, ostensibly because it irritated me to be pricked by dead leaves when I weeded around it, but more because it was a drab, green, ordinary *Ilex aquifolium* in the wrong place. My unusual variegated form which succeeded it has grown eight inches in five years (which even in the context of slow-growing holly is an impressively negative score), so it will be a little while yet before I can raid it for cutting material.

Despite this, holly is plainly a necessity for winter flower-arranging. Or should I say two hollies, for with the brave exceptions of *Ilex aquifolium* 'J.C. van Tol' and 'Pyramidalis', which are self-sufficient females, they all need a companion of the opposite sex to bear fruit. If you want berries and interesting golden-edged green foliage, grow 'Madame Briot' and 'Golden Queen' (which, perversely, is male) together. If red berries seem too commonplace, you can grow two variants of the common holly, 'Amber' (orange berries) or 'Bacciflava' (yellow). I favour 'J.C. van Tol' where there is room or inclination for only one holly, because it has masses of proper Christmassy berries, is self-fertile and does not have leaf prickles. Do not expect any of them to grow very fast. You must guard holly berries jealously from the birds in December, which means that, as it is unlikely you will have room for great sprays of holly in the freezer, you may have to net at least part of your tree.

Of course, one need not confine oneself to hollies to provide berrying material. *Cotoneaster* × *rothschildianus* is a large, semi-evergreen shrub which retains its many yellow berries until late on in winter. *Malus* 'Golden Hornet' clings on to its deep yellow crab apples, and two pyracanthas, *P. angustifolia* and *P.* × *watereri*, retain theirs – although, being pyracanthas, they cannot be wholly perfect, having a weakness for scab.

You will feel some compulsion to put ivy with your holly, no doubt. *Hedera* 'Gloire de Marengo' has an irregular silvery edging, 'Dentata Aurea' has creamy-yellow margins; but if the appetite sickens of a surfeit of variegations, 'Caenwoodiana' has dark green leaves and a stiff, upright habit. This last fact should commend it to the attention of those of us with no pretensions to expertise in 'floral art'.

I should be surprised if I were alone in finding church decoration anxious; it is such a widespread preoccupation. In our small village, there are only twenty-five regular churchgoers but more than fifty names on the flower rota: a vivid illustration of the fact that people will do almost anything for their church except attend its services. I wonder if those great minds which have grappled with the question 'Whither the Church of England?' have considered this curious paradox. Has Anglicanism been reduced to the Spiritual branch of the National Association of Flower Arrangement Societies, or is church decoration an act of worship in itself? God knows.

Allotment Rules, OK *24 June 1989*

I am in danger of becoming an allotment bore. Not surprising, really. For one thing, I live in a part of the country where allotments are falling into disuse (I appreciate that it is rather different in Tooting), so the chances of coming across anyone else who cultivates an allotment are not great. (First rule for the bore: make sure there is no one present who can cap your stories.) And the allotment has genuinely opened up a new world for me, so that I feel evangelistic about it. (Second rule: never be deterred by Doubting Thomases when you have an important message to get across.)

Until now I have had carefully to husband the limited space in my vegetable garden. It is not that small, but it has certainly restricted me in the variety of vegetables I want to grow. In order to save space I have been reduced to following the example of Boskoop nurserymen, so short of precious fertile Dutch soil that they use tape measures to gauge the correct distance before planting each rooted cutting. Now it seems the ground at my disposal is almost boundless, indeed on a clear day I can see almost from one end to the other, so my ambition can spread, like oil, to fill the space available. (Third: never resist the temptation to be smug.)

There has been a price to pay. I do not mean the £1.50 a year it costs me to rent half an allotment, which measures two chains by a rod, pole or perch, but a human cost in sweat and almost, it must be said, tears. It is my own fault. In a high-minded desire not to store up trouble for myself by chopping the roots of the perennial weeds – thistles, perennial nettles and some weeds I have never seen in my life before – into tiny pieces, all capable of growing, I decided to hand-dig the allot-ment, picking out the offending weed roots as I went. I was too proud and purist to ask to borrow a rotovator, such as all the other chaps use as a matter of course. I paid the price in burning hands and precious time spent digging when I should have been sowing. (Fourth: always take the opportunity to recount an uplifting example of the triumph of hard work over adversity.)

I do understand now why the cultivating of allotments has declined in popularity, except among city-dwellers without gardens of their own. They take over your life. Every evening now finds me wandering down the road, hoe in hand, to decapitate today's growth of the thistle which my careful hand-digging and picking so conspicuously failed to eradicate. I cannot imagine what is going to happen when the stuff is ready to be harvested. (Fifth: try to acquaint your audience with the simple pleasures of life, now so widely, and sadly, scorned.)

I suffer all the usual anxieties of new allotment holders about the far too public nature of my efforts, open to the comment of anyone still with teeth to suck. It pains me if a bean does not germinate, a courgette plant collapses, I am forced to rake stones into a conspicuous pile or the runner bean wigwam falls over, knowing that little will escape the communal eye. This anxiety is magnified a thousand-fold by the knowledge that they think that I think I know what I am doing. The purest of human pleasures may be gardening, but the sweetest is *Schadenfreude*. (Sixth: let no one persuade you that your belief that all allotment holders are beady-eyed old jossers is an outworn and patronising stereotype.)

Despite all this, I would not now readily give up my half-allotment. For one thing, it is immensely agreeable to be able to use all the seed in the packet rather than having to fold the top over and put it in my pocket for another day, which rarely comes. For another, the allotments' proximity to the cricket pitch and swings and slides means that Saturday afternoons can now be spent pleasurably by all without the usual divergences of interest putting everybody in a bad temper. (Seventh: whenever possible, cite instances of children's love of growing things and their instinctive sympathy with the natural world about them.) And best of all, there is often someone there willing to lean on their fork and listen to my views on the keeping qualities of maincrop carrots or how to confound the cabbage root fly. (Eighth: never pass up the opportunity to give someone the benefit of your knowledge and experience.)

Shock Tactics   *18 April 1992*

The trap was set while we were on holiday last year, when my defences were down. Before I knew what was up, and while still mourning the untimely death of our Clumber spaniel, I had agreed to the acquisition of a Labrador. Jet (not his real name, but one devised to protect his right to privacy) duly arrived: an intensely black puppy with a pedigree of which a duke might be proud, an amiable if adventurous nature, a mouth as soft as felt, and a destructive energy second only to that of an atom bomb. My misgivings about the impact such a lively and strong animal would have on something so fragile as a garden were almost immediately justified.

There is something faintly amusing about other people's puppy problems, I know. Ha ha, silly fools, we say, they should have thought about the consequences before the Sheraton *bonheur du jour* was chewed to wood pulp. Well, I did, which is why the dog lives in a palatial kennel outside, rather than in the house. However, as it is not thought fair or wise to leave him there all day when he is not out on a walk, he is freed from his prison at intervals to roam at will. The garden is thus at daily risk from an audacity and irresponsibility which would give a teenage joyrider the jitters.

It is an instructive sight to see an eight-month-old Labrador puppy take off from a standing start, when first let out in the morning. There is a terrible single-minded determination in his headlong progress along the grass path to the potting shed where his food is kept. His back legs are more than equal to scattering the grass behind him; so much so that hardly any now grows on the path at all. After breakfast is finished, there are several circuits of the garden which he particularly favours for the pursuit of bungy rubber toys but, like the children, he does not believe in formality of any kind – especially horticultural – and cuts corners across square flower beds without regret. These routes are stamped hard as rat runs, and the most vigorous herbaceous perennials have not yet succeeded in pushing their way through.

What is more, he has begun to cast a longing gaze over the wall which separates flower- from kitchen-garden. Temperamentally, he would count himself bounded in a nut-shell even if he were a king of infinite space. And so I am convinced it is only a matter of time before, with one enthusiastic leap, he is all over my parsnips.

I think I can bear the fact that the Algerian irises have not flowered this winter for the first time in a decade; they sputtered and gave up after prolonged suffocation under his weight, for like them Jet also particularly enjoys a south-facing position. I have even come to terms with the removal of plant labels (infuriating for a gardening writer) and the destruction of several clematis too small and fragile at this time of year to withstand his momentum. I do not even mind the way he stands on my trough of precious alpines to bark at me through the study window. What has been too much, and has precipitated a stern demand for a new one, not paid for by me, has been the systematic dismemberment of a beautiful *Camellia* 'Adolphe Audusson' which was, until he took to tearing its branches away, growing well in a large pot by the kitchen door. After a spaniel who did little more than bite the heads off daffodils, Jet has come as something of a shock.

Measures to mitigate his worst excesses have been taken, but have not so far solved the difficulty. Crossed bamboo canes placed to block his passage merely divert rather than stem, like a rock thrown into a stream. I have put down jelly-like crystals, which are off-putting to animals but not poisonous (called, snappily, 'Get Off My Garden'), round my newest and most delicate plantings, but it is beyond me to cover every flower bed. As for the paths, I may have to peg down lengths of chicken wire, to give the grass opportunity to re-establish itself.

Just recently, it has to be said, there have been indistinct, unreliable but nevertheless definite signs of improvement. Surely he is marginally slower off the blocks in the morning and slightly more careful how he crosses the flower beds? Our daily training sessions, carried out on nearby 'set-aside' which even Jet cannot substantially alter for the worse, must be paying off. A year from now, with a staid dog at my heels, I will wonder what I made such a fuss about.

Already, although I have had to accept that areas of my garden will not grow green, let alone flourish, this season, I do have the consolation of a wholeheartedly partisan, cheerful, and – after a spell lying on the irises – deliciously sun-warmed, glossy Labrador instead. Out of the dog's hearing, of course, I

can admit that I would not now be without him for all the tea
family in China – and that includes camellias.

Self-inflicted Wound   *18 September 1993*

One day, many years ago, on hearing the telephone ring while
gardening, I rammed my large garden fork with great energy
but little accuracy into the ground. One of the four tines went
deeply into my foot close to the ankle. As I stared in shaky
horror at the blood welling up through the square hole in my
Wellington boot, I was overcome with frustration at my own
carelessness. The physical discomfort was so sharpened by
mental anguish that, ever since, it has seemed to me that the
most painful wounds are those which are self-inflicted.

Recently, something of that mental pain has returned, as I consider the consequences of our selling this garden, and the house
– where we have lived for fourteen years – which goes with it.
There are some eminently practical reasons for moving but I
know that, if we had not found a house with three acres (much
of it presently a Christmas tree plantation and so virgin gardening territory), we would not now be packing up the *Just Williams*.

Although we have been very happy here, lately I have felt
that I had nowhere left to go with this comparatively small
garden. It may not always be neat, but it is so intensely cultivated that there is not now a square inch of spare soil. Had we
stayed, I would have to hack and hew and dig and delve in
order to accommodate some of the many other things I would
like to grow. Even then, I could never plant the trees I like.

The children feel confined, the dog even more so. Over the
years, my greed for planting space has shrunk the cricket
'wicket' to a thin green strip overhung by exuberant thorny
roses and overgrown shrubs. Silly mid-on now has to field in a
flower bed. No wonder the children have strongly encouraged
us to move to broader acres.

Modern life is not sympathetic to the garden-minded, unfortunately. Planning policy has disastrously promoted the turning of gardens into building plots, both in town and country, so that many houses no longer have gardens commensurate with their size. As for newly-built ones, even in rural areas there is often only garden enough for the kiddy-swing, a forsythia and a few herbs by the back door. No wonder Sunday afternoons have to be spent at Alton Towers, whether you like it or not. Acutely conscious of all that, when we finally, after years of looking, found what we wanted – a modest house but with an unusually large 'garden' – the decision to move was not so difficult.

It is only now, with the move imminent, that regret and even irritation at my lack of care have begun to undermine my sense of cheerful anticipation. Did I really think out the consequences? It will not now be me who sees *Acer griseum* grow into a fine tree, or trains the final tier of the espaliered apples. My lovely stone horse trough planted with acid-loving alpines, which took five men to lift, is cemented in place and cannot be moved. Not wishing to do the outdoor equivalent of removing the light bulbs, I must leave all my plants to someone else's possibly uncertain care. Although the garden is sliding into autumn, I find myself pottering lovingly round it, taking a wistful pleasure in plants that normally pass me by.

What has just prevented a most unworthy and unjustifiable self-pity, however, is the fact that I can be as busy as I like propagating plants by cuttings and division, and so take something of the garden with me. That is more than I can do with the house. Moreover, I can take comfort from the fact that it will not be me any more who has to try to get plants to flower in the full shade of a mature walnut tree, in a soil so dry that even Mediterranean plants have been known to flag in late summer, and in the teeth of easterly winds which come straight, unchecked, off the North Sea.

Once the shock and pain of transplantation are over, my

courage will revive and I will set about making a garden which is better, I hope, than anything I have made before, where new plants and schemes will give me no leisure to mourn those I have left behind. Then thoughts of this garden should become, like the scar on my foot, both a painless reminder of the past and a necessary spur to wiser, more thoughtful, action in the future.

Gimme Shelter   *14 November 1998*

I was, as they say, in denial. Although I had moments of rationality, generally I refused to believe the evidence of my own eyes. I did not have an 'exposed' garden, and there was an end to it. An exposed garden was one adjacent to the promenade at Bognor, or a thousand feet above sea level in the Brecon Beacons; not one at 150 feet in the gently-rolling Northamptonshire countryside. When we were buying the house, we discovered that the village had a reputation locally for being 'very windy', but I was careful not to let exaggerated hearsay and idle gossip put me off.

The blustery weather of the last few weeks has battered my conviction quite as much as it has the autumn-flowering chrysanthemums. I have been forced to accept that an exposed garden is what I have and, with that, accept another uncomfortable truth: that it is very difficult to reconcile satisfactorily the need for shelter from the wind with the equally pressing desire for a view.

You see, we have a view from the garden. Not a breathtaking view, but a comfortable, charming, commonplace English view of grassy paddock/meadow, arable fields and farm hedges on gently rising ground, topped by a flourish of deciduous woodland. This pleasing prospect lies to the south and south-west of the garden. Although in the winter the wind can veer to the north-west and, in the early spring, to the east, the prevailing winds are south-westerly. So, in the months when we spend most time outside and have leisure to contemplate our surroundings, the wind mostly comes from the direction of those fields, hedges and woodland.

Shortly after we moved to the house, five years ago, I planted a number of trees close to our boundary in the hope of encouraging wildlife and, in the long term, some shelter from the stormy blast. But my reluctance to plant the most efficient shelter-belt – that is, one composed of fast-growing, exotic, evergreen conifers – so close to open countryside, together with my desire not to block the view entirely, meant that I entrusted the task to good old native oaks, ash, birches and wild cherries. Even hollies were excluded, by reason of their dislike of our clay soil, so there was not an evergreen among them. I persuaded myself that in winter we would have the view, and in summer some protection from the south-westerlies. Some hope.

The boundary is more than one hundred yards from the house. Trees can only diminish wind speed up to a distance of ten times their height. Trees on or near the boundary will have to be at least thirty feet high to diminish at all the force of the wind near the house. This year, next year, sometime, never?

Eventually I had to admit that, in order to grow the widest possible range of plants in the garden proper, I would need to plant hedges as well to protect them. I put in two ranks, separated by lawn, to form an outer and inner stockade. The outer hedge is of hornbeam, which if trimmed in July retains most of its dead, brown leaves in winter. The inner one is of yew. One day, this double defence should ensure that the garden nearest the house remains calm when the fiercest winds are howling round the outer reaches.

Effective defences, however, will threaten the integrity of the view. The only way round this is to vary the height at which the hedges are cut in such a way as to frame the best part, or parts, of the landscape. Indeed, this may even enhance the view, as a picture frame can enhance a landscape painting. There will also have to be some judicious thinning of the boundary trees in years to come, to preserve the view without creating a wind tunnel.

Despite these flashes of forethought and lucidity, my previous denial that our garden is windy has prevented me from staking every tree sufficiently. The modern received wisdom is that, except in 'exposed positions', trees make tapered stems and sturdier root systems and are less likely to blow over in a gale if they are left unsupported or given only short, and short-term, stakes. Following this tenet has worked well for the young whips, but has been disastrous for some fruit trees put in as six-foot-high half-standards in the grass paddock beyond the hedges. Without stakes they have the greatest difficulty staying properly upright, and without protection they will become stunted and reluctant to bear good fruit. The trouble is that taking the decision to stake them heavily and plant evergreen shelter to windward will expose me to the cold blast of realism, and I don't relish the thought of that one bit.

A Load of Old Puddles   10 December 1994

A theme which runs like a never-ending stream through our gardening history is our interest in, even obsession with, water. Fashions change over the years: these days we choose to install kiddie-safe bubble millstones rather than fountains which send plumes of water three hundred feet into the air, and we have abandoned the formal pool from which emerges a winged Mercury on a stick in favour of the wildlife pond complete with frog 'beach'. But the impulse is the same.

Attractive as the idea of water must always be, the reality is not without its hidden difficulties. If your pond is made of concrete it leaks, if it is pre-formed it may be hard to make it look 'natural', if it is formal in shape wild animals drown in it while the herons pinch the fish and make the children cry, and if it is 'informal' it can soon become 'wild'. Unless unattractively barricaded, all types of pool are capable of drowning your young children or their friends.

So-called 'marginals', that is, plants which live in a few inches of water, grow thicker and stronger and quicker than any other kind of plant because they have been bred in the Pool of Hard Knocks and have learnt a trick or two about thuggery. The greater spearwort, the bogbean and the reedmaces are names to chill the conscientious gardener's blood. As for water lilies, some, like *Nymphaea alba*, know their place about as well as the head waiter of a fashionable New York restaurant.

The newly-made pond is particularly prone to problems because of the minerals which you introduce when you fill it with tap water. These stimulate extravagant green algae growth, especially as there is as yet no shade cast by water plants to restrain it. Your friends will tell you that the answer to algae is barley straw suspended in a hay net in the water – but you reply that, to your certain knowledge, such straw is not to be had, no, not even for ready money, in SW13. If there is shade, it is probably cast by deciduous trees, so leaves will float and, worse still, sink in the autumn; these soon form a carbon dioxide-rich sludge at the bottom which is more than capable of killing the fish.

The truth is that water, like fire, is a good servant but a bad master, and a bad master it often becomes. Nevertheless, I am in danger of ignoring this inescapable, ineluctable, indisputable fact. Like so many before me, I cannot withstand the siren call, or tinkling splash, of the 'water garden'. Until recently, I have always been a little superior about those of my friends who have willingly put themselves through this particular mill. But that was before Mr C. arrived one day, complete with canary-coloured digger, to push tons of topsoil around the top terrace of the garden, in a giant's game of Tonka toys. (To call him a digger driver is misleading; he is, in his way, a magician, with great sleight-of-bucket – capable of removing a heavy stone well-head cover and replacing it with minute precision, so that one disbelieves the evidence of one's own eyes.)

His eyes gleamed when I casually and vaguely pointed in the general direction of the natural depression close to the garden's boundary, and mentioned something about how nice it would be to have a 'natural' pond there – one day. Before you could say 'J.C. Bamford', he had scooped a four-foot-deep hole in the bottom of the depression to check the constituency of the soil.

Even Mr C., who is of a naturally optimistic frame of mind, had regretfully to admit that, although the soil was as difficult to get through as the omnibus edition of *Brookside*, it was not quite clayey enough to 'puddle' with the front bucket, and thereby make a watertight seal. The words 'That will mean butyl flexible liner at 50 pence a square foot, not to mention my/your extremely/pretty reasonable digging rates per hour' hung, unspoken, in the air between us. Such a course would reduce my expansive pond to an area not much larger than a puddle. Surely this was my salvation. I could make lots of noises about how disappointed I was not to be able to afford one, but really one does have to be realistic . . .

I had reckoned without everybody else's desire for me to have water in my garden. In particular, a landscape architect friend came up with the magic word 'bentonite'. Bentonite is a sodium clay which can be sandwiched between two layers of geotextile which are 'self-healing' if cut. I discovered that this stuff costs 68 pence a square foot. While I shook my head unhappily at the ruination of all my cherished hopes, she went on to say that there was a dry powder form of the stuff which, although fiddly to deal with because it swells, when wet, into an impermeable gel fifteen times its dry volume, would be much cheaper. If you want to know what happens next, watch this hole.

Give me Space

Last year, I bucked a national trend. I pulled down a conservatory adjacent to the house, and put up two greenhouses in the garden instead. You may well wonder how I could have done something so markedly against the spirit of the age. The decision may seem slightly less barmy if I admit that the conservatory was not much more than a tarted-up old lean-to greenhouse containing somewhere to sit; however, even if it had been rather grander, I might well have done the same.

It was built against a south-facing wall, as so many are; even with the glass painted with shading material, a fan going and the door and vents open, the temperature regularly topped 100 degrees Fahrenheit in the summer. As well as being uncomfortable to sit in, it was insecure. Many is the time I have woken in fright, imagining burglars moving at will downstairs, because I had forgotten to close the door to the garden. In winter the temperature inside would plummet, chilling the rest of the ground floor, despite it being heated to keep its draughty, wasteful extremities free from frost.

The 'conservatory' was crummy, but even a purpose-built one cannot easily serve two masters. Either its human occupants or their plants will be happy, but rarely both, unless a great deal of money is spent. With the exception of cacti, succulents and 'Mediterranean' plants, tender plants need lots of water (and often a humid atmosphere as well) to flourish in summer. The destructive red spider mite can only be kept at bay by keeping the humidity high. Who wants to slide along a slippery floor or have water drip down their necks while on the way to a leisurely breakfast? The central heating may well have to be kept going at night in the winter to maintain a temperature to suit the plants, yet the large area of glass will encourage heat to leak away.

So, for a conservatory to work really well – for both plants and people – it needs to be double-glazed, efficiently shaded with external blinds, heated from beneath a tiled floor, with masses of opening vents, effectively secured against burglars, and sited facing west or thereabouts. All this is perfectly possible, of course, and many owners (although by no means all) will loudly affirm that the conservatory has changed their lives for the better.

Small wonder, then, that the greenhouse has lost ground in recent years, both for aesthetic reasons and because it does not provide any additional living space. There is also a common belief that it needs the kind of constant attention which is not possible if you go out to work all day. This, however, is not necessarily so. Like the conservatory, it has benefited enormously from the development of automated equipment, originally devised for professional use. Even small greenhouses can be left all day in the summer without harm to the plants, thanks to a gadget of simple brilliance which opens vents when a predetermined temperature is reached. This works beautifully provided that the greenhouse has plenty of vents and louvres. Certainly the manufacturer of my greenhouses put in more than the standard number, when asked nicely.

It is true that, like Christmas dinner, a greenhouse needs all the trimmings. A 'windowsill propagator' or 'hot bench', 'supplementary' lighting and air-circulation fans are the Turkish delight, *marrons glacés* and Brazil nuts of the well-functioning greenhouse. But therein lies the fun of it, and the cost is a drop in the watering-can compared to fitting out a conservatory properly. The only professional equipment I have avoided is an automatic watering system: these were developed for growers of crops with identical requirements, so do not suit the very mixed collection of plants growing in an amateur greenhouse. They are worth consideration, however, by anyone who has no one to water for them when they are away from home.

Any amount of bubble polythene or fleece insulation can be used in the winter to reduce the heating cost, without fastidious eyebrows being raised. This is because the greenhouse is seen as purely utilitarian. Unlike the conservatory, it is not there to suit the garden owner. Yet, paradoxically, it will probably provide as many happy hours as a conservatory (although of a different nature), especially at this time of year. Sowing a small tray with seed, removing a few dying leaves, taking the time to observe the budding-up of forced flowers, breathing in freesia scent: it is these small pleasures which make a greenhouse such a solid joy and lasting treasure. What is more, and in marked contrast to the conservatory, it provides a respectable retreat from the rest of humanity. In other words, a living space.

2

Trees and Shrubs

Like displaced Russian noblemen driving Parisian taxis to pay the rent after 1917, magnolias cannot quite hide their sense of innate superiority. They are émigrés forced to come down in the world, and they do not much care for it.

While magnolias were natives of this land mass in the Pliocene age, some million years ago, climatic changes chased them out of these progressively colder regions and left them stranded in North America and east Asia. Their lineage, therefore, is more ancient than any to be found in *Debrett* or the *Almanach de Gotha*. As a family they are the most distinguished and glamorous of spring-flowering shrubs and trees, but the wants of many of them are exacting and expansive. They require the wide spaces of a large garden in which to spread their branches or, better still, the half-shade afforded by an established woodland, and a generation must necessarily pass before many of them consent to unfold their flowers freely.

They have a preference for a soil which is rich and open, deep and peaty, and a distaste for being uprooted which would do credit to an aristocratic family fallen on hard times.

The most gorgeous of the evergreen magnolias, *M. grandiflora*, requires full sun and shelter and grows to fifty feet, which means that it needs, except in most favoured localities, the backing of the wall of a three-storied south-facing mansion with acres of brickwork between the windows for its huge glossy-green leaves to fill. The sparsely-produced summer flowers may be the size of soup plates and are headily scented, but they will take ten years or more to appear in any quantity, and rarely do so near enough to the ground to be smelled in comfort and safety. Magnolias in general are in no hurry to accommodate themselves to the impatient demands of bourgeois convenience – here today, and somewhere else in two years. In short, they are out of their time.

It was not always so. In those now far-off days between the wars gilded in the memories of old men, the first flowers of magnolias grown from seed collected in south-west China were anxiously but patiently awaited for twenty years or more. The City luncheons given by Lionel de Rothschild in the 1920s buzzed with talk not of the Big Bang, or the possibility of single-figure yields on gilt-edged stock, but of how many wild-collected seeds of *M. rostrata* had been successfully germinated. The plant hunters George Forrest and Reginald Farrer sent back ecstatic reports of what a wonderful plant it was, but even experienced botanists can get it wrong, and they had mistaken *rostrata*, with its small and, for a magnolia, relatively insignificant flower for the showier *campbellii*, so that the wait from 1918 to 1935, when the first one flowered in England, was more or less in vain. There was room in the country gardens of horticulturally-inclined City gents for the larger and more difficult magnolias, the time for such preoccupations, and the staff to look after them.

Today, in big gardens in Cornwall and Devon and the

Atlantic seaboard of Scotland, *M. campbellii* should be flowering now, its rich-pink flowers fluttering near the tops of fifty-foot trees like great pink birds alighting on the bare, leafless branches. In Sussex, in a few weeks, the pure white flowers of *M. sieboldii* will open and exhale their heavy, lemony scent. Here they are happy. Thoughts of these exotic magnolias, however, only serve to point up the inadequacies of my own circumstances and prospects, gardening as I do in the cold and windy East Midlands. What hope have I, in my average-sized garden, already crammed with plants, which is limey, prone to drought and exposed to easterly winds and early morning sun, of succeeding with these patrician plants, whose flowering time coincides with the most damaging spring frosts, when the sun comes out to scorch the frozen flowers and turn their white goblets into rotting brown rags?

There are some members of the magnolia clan, fortunately, which extend a tentative hand of friendship to those not born with the advantages of space or temperate climate. The approach may be a little condescending, and certainly comes more readily from the poorer relations of the family, but there is no such thing as an ugly magnolia, so I must seize any opportunity I have to ingratiate myself.

The most accommodating offspring is the hybrid *soulangiana*, which lords it over the humble forsythias and flowering currants in many an adventurous suburbanite's garden. There is also *stellata*, with many strap-shaped petals and a certain star quality. Both of these are relatively easy and reliable, flower early in their lives and will put up with the cakes and ale of ordinary soil, even if they do hanker after smoked salmon and champagne in the form of the rich peaty soil available in southern and western gardens. These resolutely refuse to grow into stately trees, preferring to remain as more lowly shrubs. I shall shortly be planting *M. × highdownensis* in my garden and, if I had acid soil, I would plant its parent, *M. wilsonii*; these will grow against a medium-sized wall, and do not open their cup-shaped flowers till May. The modern City man in Clapham or Islington should find room for the deep pink 'Leonard Messel', named after the banker who lived at Nymans in Sussex.

Gardeners who do not, however, feel up to the strain of putting these lofty and rather tricky guests at their ease might like to go in April to gaze at them in their public finery, at Nymans or Leonardslee in Sussex, at Lanhydrock or Trewithen in Cornwall, at Crarae Glen in Argyll – gardens which, like the magnolias, are relics of a more spacious age.

Two Nations *25 April 1987*

'In the home paddock, where the grass had grown three feet high, we ran into thousands of famished leeches, and almost naked coolies arrived in camp with arms and legs striped like a Bengal tiger. The headman was so covered in blood, which streamed down his face, as to be almost unrecognisable.' That is Frank Kingdon Ward describing a commonplace incident during a plant-hunting expedition to the mountains of Assam and northern Burma in the late 1920s in search of rhododendrons and primulas. Leeches were one of the relatively minor hazards faced by a man who conquered a mortal fear of both snakes and heights to collect some of the finest rhododendrons ever introduced into this country.

Considering the lengths to which he (and other plant-hunters such as Sir Joseph Hooker, Robert Fortune, George Forrest, Reginald Farrer and 'Chinese' Wilson) went in order to collect them, it seems, to say the least, ungrateful of many gardeners to be so dismissive of rhododendrons.

We are two nations separated by a common interest in gardening. Cultivators of alkaline soils recognise but do not really understand the language of those who have a pH of 6.5 or less. This lack of understanding leads to prejudice. 'How I dislike rhododendrons – great, coarse things – I'm so glad I can't grow

31

them.' As with Guinness, we do not like them because we have never tried them. Thus a whole order of plants, which rivals in size and splendour even the orchids and the roses, is consigned to the outer darkness of our indifference.

If we think of rhododendrons, which is rarely, it is to *R. ponticum* that our minds turn, then quickly turn away. This shrub (which masses along the margins of every dark driveway in Scotland) has sickly mauve flowers above dank, undistinguished, evergreen leaves, and imperialistic tendencies of which Tamerlane might have been proud. But then, *R. ponticum* is not popular these days with anyone.

It may be useless to point out to the unbeliever the impact of a wellgrown 'Beauty of Littleworth' with its huge white flowers, or the perfect, dark crimson, waxy, elongated bells of *R. cinnabarinum* var. *roylei* 'Magnificum'. Perhaps it is also fruitless to mention the great impression made by a fifteen-foot specimen of the blood-red *R. thomsoni* in a woodland glade, or to dwell on the enormous size of the leaves of *R. sino-grande*. Presumably even talk of the deciduous azalea *R. luteum*, with its deep yellow flowers blanketing the bush in May, and a sweet scent so powerful that one can catch a breath of it from inside a closed car, will fall on deaf ears.

If one can be persuaded to like them, there are at least two rhododendrons which, surprisingly, will grow on limey soil. One is 'Cunningham's White', a fine and extremely hardy but uninvasive *ponticum* hybrid which grows up to ten feet tall and has masses of flowers which start faintly mauve but become pure white save for a little yellow speckling in the throat. This can be used as covert-planting for those with a shoot to maintain who wish for something other than the ubiquitous snowberry to please the pheasants. The other is a small alpine shrub with dark pink flowers called *R. hirsutum*.

There exists an ericaceous planting compost which will satisfy the craving for acid conditions which is the hallmark of the Ericaceae, the family name of rhododendrons and heathers. There is no reason, for example, why we who live on alkaline soil should not grow the small, neat, compact 'yaks' (the *cognoscenti*'s word for the forms and hybrids of *R. yakushimanum*) or *R. moupinense* in large pots or tubs. Separated by a *cordon sanitaire* from alkaline soil, watered with rainwater rather than with tap water (which, in East Northamptonshire at least, is as hard as an ungrateful heart), there is no reason why the leaves should ever yellow. If they do, spraying with sequestered iron, which temporarily unlocks the grip of calcium on iron in an alkaline soil so that it becomes available to the plant, will soon put matters to rights.

Even if you cannot or do not wish to grow rhododendrons or azaleas, it seems a pity never to take the opportunity to look at them. I am taking my own advice, for once, and spending some days next week on the west coast of Scotland in order to see Scottish glens transformed into Himalayan gorges. The unconvinced might like to begin with a tour of the Savill Garden in Windsor Great Park or the Royal Horticultural Society's gardens at Wisley in Surrey, or one of the large rhododendron gardens of Sussex or Cornwall. Wherever you go, there will be plenty of Himalayan introductions to see and even, perhaps, admire – but no leeches.

Social Climbing   *16 November 1991*

It may not have been the most commercial enterprise ever to be undertaken by Messrs Jardine and Matheson, East India merchants, nor even the one for which the company will best be remembered, but their introduction of the Japanese vine *Vitis coignetiae* from the forests of Hokkaido to Anthony Waterer's famous tree and shrub nursery at Knap Hill in Surrey in about 1875 should have earned them the eternal gratitude of gardeners.

In my garden, this immensely vigorous ornamental vine

has scrambled, with the initial help of large-gauge plastic netting, more than twenty feet up the smooth trunk of our Tree of Heaven (*Ailanthus altissima*). In late October and early November, the large leaves turn a dozen shades of scarlet and orange, hanging on short laterals away from the tree, so that they glow when the sun shines through them. The sight profoundly alters the appearance of a rather dull corner of the garden and entirely eclipses a group of squat shrubs seated below. We take our autumnal tone from this climber: if the colour is good and the leaves hang on well, we remain cheerful at least until Christmas; if wind or frost make the leaves drop too quickly, spring suddenly seems a long way off.

As a matter of course, I planted this vine against a tree. Any mature tree in this garden gets the same treatment, so that trunks sprout like the experimental stubble on a teenager's chin. But while the vine has plainly found its home from home, not all climbers look so well.

It is an axiom of modern gardening that plants are naturally sociable, and they like to huddle together. By clothing every tree I have gone along with this, although in fact it ain't necessarily so. Plants differ in characteristics and wants as much as the occupants of an Underground carriage, so we cannot blame them if some react as we would if forced to huddle too close to other people in the Tube.

Roses, *pace* Gertrude Jekyll, are a decidedly mixed blessing. They are too stiff, once the older wood lignifies, to give that sinuousness which is appealing in this situation. In addition, in the wild they scramble through scrub and thicket to reach the light, and natural pruning consists of a random shedding of dead wood. In the context of a garden, pruning for a good flowering cannot be so simple or haphazard. I have, for example, the pretty pink-white Noisette rose 'Madame Alfred Carrière', climbing up and through an old Victoria plum tree. The most vigorous shoots of this rose extend beyond the higher plum branches by several feet. An ordinary ladder or stepladder is useless for pruning purposes, therefore. If I cut from below, I have the problem of somehow pulling down dying branches without harming the plum tree. Leaving them there is unthinkable.

A better bet, in this situation, might be a clematis, but even then I should have to find the right one. The easiest to manage are those which naturally benefit from being cut down to a foot or so from the ground in the late winter. However, of these, only a few are vigorous enough to make a suitable impact without cleaving to the trunk like a frail Victorian maiden clinging to a burly soldier who is anxious to be off to the wars. This means one's choice is restricted to a few very strong-growing named varieties like 'Perle d'Azur' and 'Ville de Lyon', or one of the species clematis, like *Clematis orientalis*.

Fortunately, there is a small group of climbers which are more suitable for large trees, and even less bother. The so-called 'self-clingers' are those plants, like *Hydrangea petiolaris* or *Schizophragma integrifolium*, which can stick like limpets to a rough surface like fissured bark, and will happily take some shade.

In addition, there are the annual or herbaceous climbers which die down naturally in the winter. A striking effect can be achieved if the scarlet 'flame flower', *Tropaeolum speciosum*, is planted to climb a close-knit evergreen, like a yew tree. It will not do at all in my thirsty alkaline soil, but I can achieve a similar impact with the orange-flowered half-hardy climber *Eccremocarpus scaber*. It is vulgar in appearance, shameless in its determination to get on, yet unable to support itself by its own efforts. A true social climber, in fact.

The Good Companions   *23 August 1997*

Who would buy a hemerocallis or a dregea, a vitex or a schizo-stylis on spec? Some plants are as badly served by their names as people. Clematis is the unluckiest, for there are two ways of pronouncing it, and neither could be called euphonious. To be fair, the botanical name is preferable to 'Virgin's bower', championed by William Robinson in the nineteenth century. Happily, the epithet never gained universal currency, even in the days when virgins and bowers were two-a-penny. Nowadays, um, ah, bowers are thin on the ground, in any event.

We should not let a name put us off the 'Queen of Climbers' (another unappealing expression), for it is a wonderfully diverse and useful group of plants. For most people, 'clematis' means the striking 'large-flowered hybrids'. These are the crowd-pleasers, and there are times and situations when only they will do. However, for me, as for Robinson himself, the true beauty of the family resides in the smaller-flowered species, and their related garden varieties: the alpinas, macropetalas, texensis and viticellas.

Clematis have probably never been more popular, presumably because they can be squeezed even into tiny gardens, they will grow in pots, and they make excellent cut flowers. But they could be bigger still. Their appeal is limited by the fact that they are not straightforward to grow.To begin with, their pruning is perceived as complicated, although it is simplicity itself – once the difference between 'new' and 'old' wood is understood. Only with correct pruning, however, can a clematis give of its best rather than hanging like a tangled ball of wool from one wire on a wall, or producing a few reluctant flowers on the top of one weak stem.

Furthermore, some of the flashiest hybrids are not very hardy; some are slow to get going; they occasionally succumb to 'wilt' and turn to slimy string before your eyes; and most need a lot of water. They cannot be plonked in the ground and forgotten, as many other plants can.

Above all, to get the best out of clematis, the gardener must have an idea of where best to place them in the garden. For they are companion plants, *par excellence*. In the wild they clamber over other plants, twining their long leaf-stalks round any support. They do not seem to want to live on their own. They are like a rich old widow, prepared to put up with gossipy nonsense from a paid companion, just to keep loneliness at bay. The gardener can use this undiscerning sociability to her advantage, dressing up the dowdiest out-of-flower lilac or evergreen shrub in clematis finery. Alternatively, those with bluey-pink, blue or purple flowers can be used to complement the flowers of other genera, like roses. You only need to know when particular varieties of clematis flower, when they can be cut back, and where to site them.

The gardener must also cultivate the soft hands and patience of an opening batsman. The clematis's companionability is combined with a fragility which can be heartbreaking. You feel a complete fiend when you accidentally break a delicate growth while attempting to guide it through another plant, even if your guilt is immediately tempered by irritation at the family's wilfulness.

So why do we bother with them? Because, when grown well, in the right place, and in full bloom, there is no more sumptuous flower than 'Royal Velours', no more elegant charmer than 'Duchess of Albany' with its upturned lily-tulip flowers, no more delicate beauty than the bell-shaped *rehderiana* nor more spectacular scene-stealer than 'Jackmanii Superba'. What is more, there are so many of them that they can become a true, all-embracing enthusiasm, comparable with a passion for roses or irises, alpines or fuchsias, chrysanthemums or dahlias.

Clematis enthusiasts, or 'clematarians', are as sociable as the flower they cultivate. Although only founded in 1991, the British Clematis Society already has a membership of 1,200. Perhaps because of its youth and the type of people clematis attract, the society has a reputation for being unstuffy and welcoming. Members receive fact sheets, a well-produced yearly bulletin and frequent newsletters. Clematarians meet often for seminars and lectures, and for visits to nurseries and each other's gardens. They organise plant sales, are present at most big flower shows, and exchange clematis seed and plants. So if your garden is, or you would like it to be, filled with trellis, tree stumps, pergolas and even, perhaps, bowers, all hidden beneath cat's-cradles of clematis, I suggest you join them.*

In Pursuit of Daphne *18 January 1992*

I usually come across it at the beginning of the fifth drive, when we have finished beating the fields and just gone into the wood. As we spread ourselves out in a line to wait for the whistle I am careful where I tread, for this is the only place in the whole vast wood where *Daphne laureola* grows.

On each of the many Boxing Days when I have stood here, waiting for the guns to reunite themselves with disobedient Labradors, stop arguing about the tactics employed by Robert E. Lee at the Second Battle of Bull Run and get themselves into position, some early flowers on these gaunt, evergreen sub-shrubs have gleamed dully in the subdued light. The warmth and shelter of the woods, which lure the pheasants in from the windy fields, must coax these plants into precocious flower. At the time of the last shoot of the season, in very late January, the clusters of yellow-green, four-lobed tubular flowers are showing fully from the circlets of glossy, dark green leaves grouped at the top of almost leafless three-foot stems.

The other beaters tease me for my interest in these obviously dull plants, although they also avoid wantonly treading on them. They call them 'rhododendrums' for, like most country-men, they are rather better at naming animals than flowers. (The same cannot be said of the guns, who not only rarely know hellebore from bugle but sometimes have difficulty

*British Clematis Society, 4 Springfield, Lightwater, Surrey GU18 5XP; www.britishclematis.org.uk

telling flying hen from cock pheasant.) It would be strange indeed to find rhodies growing on the edge of the chalk Chilterns in Oxfordshire, but this deciduous, predominantly beech woodland provides just the conditions the spurge laurel seems to require.

It is one of those plants referred to by the wild-flower books as 'widespread and common in southern calcareous woods', which means you can count yourself very lucky to find it. It is, incidentally, neither spurge nor laurel, although I suppose if the wood spurge could breed with the cherry laurel the result would look very similar.

It is not its uncommonness which appeals particularly, however, but the fact that it flowers so soon after the shortest day. When you are leaning against a leafless tree, sucking a humbug, talking to the dogs and waiting for something to happen, even the unfolding leaves of the honeysuckle will catch the eye, and a flowering shrub is an event to be remarked upon. Like most winter-flowering plants, it is scented; the pity is that the fragrance is strongest in the evening, when we have hushed our rude noise and agitation and the woods have once more fallen silent save for the occasional croak and clatter of returning pheasants.

I have long since given up looking for the other native daphne, *Daphne mezereum*, in that wood. Even the books admit that the mezereon is now rare and confined to a very few calcareous woods, the whereabouts of which they are too cautious to reveal. However, it is common enough in commerce and grows easily in the garden. It comes in two flower forms, unsubtle purple-pink and paper-white. It is deciduous, unlike the spurge laurel, and the scented flowers are borne in late January, tight against the naked, upright branches. This aspect moved William Cowper to write of the mezereon:

> Though leafless, well attired and thick beset,
> With blushing wreaths investing every spray.

The pink form bears spherical red berries while the harder-to-find white form, 'Album', has yellow ones. All these berries are poisonous. The mezereon has an irritating habit of dying suddenly, after just a few years, from virus, but there are usually one or two seedlings growing in the garden from seed spread about by the birds.

The spurge laurel lives longer than the mezereon, and it is one of the few plants which will happily stand the drip from trees. It is unusual among daphnes in having black rather than red berries. *Daphne laureola philippi* is the best form to grow because it is shorter, so lacks the gauntness of the species.

It is one of those plants, and they are surprisingly common, which are fussy as to their circumstances in the wild but adapt to a variety of soil conditions in the garden. I cannot pretend that the flowers are showy, but this daphne deserves notice for daring to flower in the middle of winter – and for providing something to look at while we wait for the hunter-gatherers-for-the-day to stop gossiping and find their pegs.

Cypress Gang 13 October 1990

When I was in America some years ago, I recall seeing a sign outside a McDonald's with the legend 'Over 10 billion burgers sold'. I used to wonder how big a pile such a monumental number would make if they were laid one on top of the other – surely as tall as the Empire State Building, or the CN Tower in Toronto if you added in the gherkin. It certainly seemed a good advertising ploy, appealing as it did to many people's desire to be one of a huge crowd. I can only hope that the large garden-centre chains in this country do not take to putting up banners at their gates which read 'Over 10 billion Leyland cypresses sold'. For it would ensure that the few who have resisted this conifer until now could no longer hold the line, and its dominance as *the* evergreen hedging plant in British gardens would be complete.

The Leyland cypress, × *Cupressocyparis leylandii*, is not a true species but an inter-generic hybrid between the Monterey cypress, *Cupressus macrocarpa*, and the handsome false cypress, *Chamaecyparis nootkatensis*. The name 'Leyland' comes from C.J. Leyland of Haggerston Hall, Northumberland, where the first seedlings, which were actually raised at Leighton Hall near Welshpool, were sent in the early 1890s. This tree did not come into general commerce until the inter-war period, but when it was found to combine the hardiness of the Nootka cypress with the speed of growth of the Monterey cypress its future was assured. Indeed, it is the fastest-growing conifer in this country, putting on as much as three feet a year. It is also tolerant of close trimming and a wide variety of soil conditions, which is why it is universally popular as a hedging plant.

For some years, however, gardeners and tree experts have been expressing reservations about it which have nothing to do with its commonness. (The yew is a very common hedging plant but, rightly, does not come under the same fire.) The Leyland's very speed of growth has come to be seen as a grave disadvantage. Planted to define a boundary, it soon becomes a too-solid barrier which needs clipping three times a year – in May, July and September. That means a lot of afternoons spent risking electrocution with the hedge trimmer. Not everyone has the energy or nerve for that.

In the gardens of houses which change hands frequently, any interregnum gives the Leyland cypress the chance to tear away. They are the Teddy Boys of the horticultural world, kind to their mothers, no doubt, but inclined to get out of hand when in a gang. Although they can be clipped closely, you cannot cut them hard back, for they will not break from the old wood the same way as yew will.

The Leyland cypress does not even make a very neat hedge. The arrangement of branchlets is quite lax, so that, except just after clipping, it can look a little scruffy. There are far neater hedgers among its cousins the false cypresses, *Chamaecyparis lawsoniana*. It has altogether too exotic a look for country gardens. The golden form, 'Castlewellan', is hardly more of a laggard and has the disadvantage of yellow foliage, which is dingy in winter and makes a restless backcloth for plants.

I am not a Leyland hater, but I believe that it should be grown only as a specimen in a large garden and given plenty of room to make its characteristic columnar shape, which will be sixty feet tall after twenty-five years. Although I have an inherited one in my garden which is quite handsome, I would not go so far as to plant one myself.

Because the Leyland cypress has only been around for a hundred years, no one knows quite what its ultimate height will be. However, informed guesses have been made. In deference to their customers' peace of mind, therefore, the garden centres really should put up signs which announce that the Leyland cypress will only stop growing when it has reached a hundred and fifty feet or so in height. Or about six hundred Big Macs.

Give a Stork a Home   17 April 1993

I was alarmed to discover recently that the sales of trees and even shrubs are on the decline, whereas those of annuals and herbaceous plants have never been higher. The reason is that now, more than ever, people are planting for the present rather than the future, and even a shrub may take inconveniently long to come into its own. For all the lip-service paid to the importance of planting trees, the idea of posterity no longer has much resonance for us.

Take the fate of the cedar of Lebanon. *Cedrus libani* is scarcely planted these days outside public gardens and arboreta; it is only listed in the catalogues of a handful of specialist nurseries. Yet there are, especially in the south of England, enough mature (and, it must be said, over-mature) specimens of this imposing, spreading, distinctively horizontally-branched large tree in gardens for us all to recognise one when we see it. These were planted before the phrase 'instant gardening' had any meaning.

Cedrus libani was introduced into this country in the middle of the seventeenth century, but the majority now extant have been planted in the past two centuries – as often, I suspect, for their biblical associations as for their majestic presence. The timber, so durable and fragrant, was used by King Solomon to line the inside walls of the Temple in Jerusalem and also his private palace, called the 'House of the Forest of Lebanon'. As the Psalmist has it,

> The trees of the Lord are full of sap,
> The Cedars of Lebanon which he hath planted
> Where the birds make their nests:
> As for the stork, the fir trees are her house.

As these 'fir trees' were already rare in the mountains of Lebanon by the time of their introduction to Britain, it must have seemed an act of piety to keep the species from possible extinction. The best specimens are in the warmer south and west. They seem happiest in a deep, well-drained but moist soil, but are not picky, for they do well at Kew Gardens, which could never be accused of having a really good soil.

In time, this handsome tree came to exemplify a certain kind of old-established garden, of Wodehousian timelessness. This garden was not necessarily big; I remember, as a child, playing on a swing suspended from the branch of a cedar, and our flower garden measured less than an acre. But it was a garden planted for permanence, a permanence so obvious then as not to need expressing. It is the idea of natural continuity which has been fatally undermined in the past thirty years. Mobility of labour, demanded by many-sited businesses and professions, is knocking the staying power out of the gardening middle classes and forcing them to join the ranks of the 'bedders-out'. Why plant a tree, any kind of tree, let alone one which does not mature for decades, if you know that the longed-for promotion means relocation to Peterborough?

Ironically, there is a cedar now widely planted which, in its native Atlas mountains, grows as tall as the cedar of Lebanon, and sometimes even acquires the same distinctive flat-topped and expansive look. What gives *Cedrus atlantica* 'Glauca' its 'garden centre appeal' is its attractive bright blue-grey needles, helped by the deafening silence maintained by nurserymen about its eventual size of eighty feet by thirty.

Most will not develop fully, of course, for they will be cut down long before their time. Ignorant of their potential as gardeners seem to be, they have planted blue Atlas cedars at the corners of ten thousand patios, and these cedars are on the way to obscuring the view from ten thousand picture windows. They are designer trees, like designer pets, garden accessories to be admired but discarded when they become inconvenient; they will never know a rickety and individualistic old age.

Cedars were hit, along with all other trees, by the ferocious gales of 1987 and 1990; unlike other trees, however, there seems to have been little impetus to replant them. Certainly in private gardens which can still boast a substantial tree it is rare to see a small specimen planted nearby against the day when it finally dies. (The situation is most hopeful in public gardens, not surprisingly; the National Trust, for example, has a programme of planned replacement.) Yet with careful siting many gardens, particularly country ones (as this tree is sensitive to atmospheric pollution), could accommodate one at the far end of the lawn. After all, the roots do not quest inordinately after underground water supplies, like those of the willow or poplar both common in medium-sized gardens, and they are no more prone to windblow than other conifers.

Moreover, if you plant one of these truly distinguished trees, not only will you have done your part for posterity, but you might just encourage a stork to come to nest in its branches. Well, you never know.

Woodman, Spare that Tree?   16 October 1993

The first thing we decided on moving to a new garden was that we would cut down some of the trees in it. The second thing we decided was that we would not. The third thing we decided was that we would . . .

A certain hesitancy is inevitable, I suppose, considering how important trees are in a garden, and how long it takes them to transform themselves from spindly sticks to what garden designers call a 'feature'. But it goes deeper than that.

A tree is more than a mere concatenation of xylem and phloem, chloroplast and cambium, more than a shrub that did not know when to stop. It is more even than a beautiful natural object. A tree is a potent symbol of permanence, of stability, which we lack as much in our gardens as in our lives. Nothing, but nothing, gives a garden a more established and mature look than a tree, however undistinguished, which is more than twenty years old.

Trees can also be repositories of both history and folklore. You only have to think of Robin Hood's oak in Sherwood Forest, or the yew in the churchyard at Selborne which Gilbert White knew. It is not simply the enormous age they are capable of achieving, but the associations they gather on the way. It is not too fanciful to say that deeply buried in our psyches is a memory of tree-worship.

The predisposition should always be to retain a tree if possible, hence our twittering and dithering. The garden which we have bought was laid out in its present form about forty years ago, or so I guess from the number and size of the ornamental cherries in it. They must have seemed eminently suitable, for they are reasonably unfussy as to soil, do not attain the size of a beech or ash, and have flowers which bloom in the spring, tra-la.

We have inherited the most popular, and overrated, of them all. The combination of copper-coloured young leaves and frilly pink, double flowers of *Prunus* 'Kanzan' affects me like grit in the eye. Unlike a nearby *P. sargentii*, which has the grace to turn to fiery red and yellow in autumn, the leaves of 'Kanzan' stay as green as grass almost until they fall. Its 'charms' are therefore short-lived.

We also have a Lawson cypress, which I suspect was planted many years ago as a 'dwarf or slow-growing conifer', for nurserymen were even less precise then than now about how tall conifers will grow. I cannot think why else anyone would put a large tree (which has already grown more than thirty feet tall) within twenty feet of the house. It not only blocks a great deal of light from the terrace and house, but masks a shapely, spreading 'Bramley' apple tree which is growing beyond it. Even though it is handsome when considered in isolation, no one could blame us for removing it.

As for the 'Kanzan', I doubt if I will find it difficult to cut down, despite my natural reluctance to get rid of a perfectly healthy tree. I have only to think of the next twenty or thirty springs spent looking at those silly tutus of flowers to stiffen the sinews and summon up the blood.

As gardens are constantly changing, nothing should be sacred in them. Nevertheless, as trees are the clearest expression of permanence in the garden, we surely have a duty both to think very hard before removing one and, moreover, to plant at least as many as we cut down – provided that there is genuinely space for them to grow and spread comfortably and conveniently.

We must take care to plant appropriate and attractive species which are in scale with their surroundings (this does not

necessarily mean small) and also suited to aspect and soil conditions. Otherwise, the trees we envisaged as permanent features, like those under the shadow of the axe in this garden,
will last no longer than our own tenure of the garden.

*The next article was written shortly after the Great Storm of 1987,
and the one following it a decade later.*

Losing Old Friends   21 November 1987

Not since the chance introduction of *Ceratocystis ulmi* in a consignment of logs from Canada in the late 1960s has such a dev-
astating blow been dealt to the tree population of England –
and the human population which owns and cultivates it.
Often, while talking to owners and gardeners of medium- and
large-sized gardens in the south and east, I have been struck by
how many refer to their fallen trees as old and much-loved
friends. Those not still in a state of shock are in mourning for a
great and unencompassable loss.

There have been times in the last weeks when I have felt like
a brash hack jamming my foot in the door to intrude on a
private grief: 'How do you feel now that your life's work, and
that of your father and grandfather before you, has been
destroyed in a couple of hours?' I have felt ashamed of the
smug calm with which those of us in the Midlands greeted the
news of the hurricane. When you have lost a single branch off
an old plum tree, it is hard to imagine what it is like to lose
more than two thousand trees, as they have at Leonardslee in
Sussex.

The latest estimates of the number of trees which have fallen
as a result of the storm in gardens, parks and woodland stand
at about five million, but these will be revised upwards as
many trees turn out to be unsafe and need to be removed. That
number is somewhere between one-quarter and one-half of

those thought to have been lost in the Dutch elm disease epidemic, it is true, but such a comparison should not be used to undervalue the scale of this disaster, particularly to those people who have sustained loss. For one thing, the geographical range of devastation in the case of the elms was much wider, and for another, the shock was more diffused. After the first elms contracted the fatal disease, there was a decent interval when we could prepare for the inevitable. Injections of fungicide were used so that patients rallied for a time, and we felt we were doing something constructive. We could learn gradually to accept the idea of eventual death. In a seemly and orderly way the trees were cut down and cremated, not torn out brutally and precipitately by the roots.

There is much talk about the importance of replanting (it is easy to be bracing when one has not lost anything), but in the larger landscape and woodland gardens of Sussex and Kent it will be many weeks or months before the débris is removed and the damage assessed, and replanting is at the moment an irrelevance. What really matters now is how to make safe those trees still standing and remove the fallen.

After the clearing-up, planting will be the easy part, although what to replant is exercising many. As hardly any species of tree seemed everywhere able to withstand the hurricane, it is pointless to plant only known wind-resistant trees: Wellingtonia, Monterey pine, swamp cypress, western hemlock and the like. Deciding not to plant a beech just because it is shallow-rooted would seem rather self-denying.

In urban and suburban gardens forest trees are inappropriate anyway, so this is a splendid opportunity to plant something more suitable than the chestnut or plane which used entirely to shade the back garden. I anticipate a great run on stocks of ornamental crabs, cherries, rowans, maples, and thorns.

The cost of both tree surgery and tree removal, as well as of replanting, even in smaller gardens, is a considerable anxiety to many. I do feel particularly sorry for the owners of private gardens which open for charity under the National Gardens Scheme. All but the most severely affected are manfully agreeing to open their gardens next year, despite being dependent largely on their own resources to get the damage repaired and begin the replanting. With the exception of the public money being allocated by the Countryside Commission for distribution by local authorities, private gardens, unless historically important, will not be entitled to receive grant aid.

For all gardens which open to the public, the future support, moral or financial, of visitors will be vital. Deciding not to go and visit a large woodland garden in Sussex, for example, because one imagines there will be much less to see there for some years, will damage its chances of recovery. The last thing the owners of gardens need in their mourning is to be shunned by their embarrassed friends. Next season, they will provide the teas if we can provide the sympathy.

Talking Trees 18 October 1997

Everyone has a horror story to tell about the gales of 16 October 1987. In our garden, I seem to remember, a branch fell off a plum tree. But then, I live in the Midlands. In the south-east of England, the Great Storm is engraved on the communal psyche. As winds gusted at speeds of up to 110 mph in the early hours of that morning, there can have been few people who did not lie awake thoroughly unnerved by what was happening outside. Although in the event casualties were surprisingly light, damage to property was enormous.

But it was the effect that the winds had on trees, still in leaf and in soil loosened by ten days of heavy rain, which will be longest remembered. Fifteen million of them are said to have been uprooted or shattered, in city and in countryside, in a line from Hampshire north-east to Norfolk. Many were young trees in plantations, but most of the tallest, and many of the rarest, ornamental trees in parks and gardens went down that morning. Least affected appear to have been trees in ancient woodland.

I remember spending time in a number of large gardens a few days after the Storm, meeting shocked owners and gardeners. (A Sussex GP told me that her surgery was full the next week of people exhibiting the symptoms of shock normally seen after a car accident.) Some had seen the work of several generations destroyed in a night. For many people the losses were truly heartbreaking, and the thought of the expense and botheration terrifying. Copious tears must have been shed not only for the irreplaceable lost giants, the so-called 'champion' trees, but also simply for local, familiar landmarks. All gardeners, wherever they lived, could sympathise.

For a short moment, trees had ceased to be powerful symbols of permanence and continuity and had become unpredictable, frightening – and disturbingly mortal. Gardens had changed out of recognition in a few hours. One immediate response to this was a lot of talk (far too much, in my view, and certainly not from garden owners themselves) about the need quickly to fell damaged trees, clear fallen timber and replant.

It was heartening how many tales emerged of courage, fortitude and a 'spirit of the Blitz' neighbourliness, as people strove to assess the size of the problem and deal with the practical difficulties these fallen trees posed. Heartening, too, has been the way that, ten years on, those who own and manage parks and gardens open to the public seem to have picked themselves up, dusted themselves off, and learned to find some good things to say about the Storm. In places, it is admitted that not enough long-term planning had preceded it, so that many ornamental trees had been allowed to become overmature, or were unsuitable for the soil or position anyway.

The Storm provided opportunities for new and sensibly staggered plantings, and in the meantime there was a ready-made ecological beanfeast for insects and fungi everywhere timber could be left to lie. The Storm also, perforce, gave the chance for some adventurous propagation: many rare trees were successfully propagated at what was often the 'wrong' time of year. And the Storm taught essential lessons about the ecology of trees: that a fallen tree is not necessarily a dead one, and that some will survive, if pulled back into their holes.

In short, although some precious garden views and vistas have disappeared, perhaps for generations, generally the Great Storm is now seen as almost beneficent – by those gardens which were the beneficiaries of appeals or eligible for public grants, that is. So much has been learned, so much rethought. Even taking into account a hefty dose of *ex post facto* rational-isation, the point is a good one.

For those with gardens which are not open to the public commercially, it is a different story. How can the Great Storm be a good thing, when you have to depend on your own resources and the kindness of friends, and where the precious atmosphere of maturity provided by trees has been lost for a lifetime?

It is hard, if not impossible, in those circumstances, to be philosophical and to admit that ornamental plantings, because they are essentially artificial in their inception and mainten-ance, will always be vulnerable to extreme climatic incidents, be they heavy snowfalls, drought or high winds. It is the price we pay for doing what we want. Garden-owners in the south-east will snort at that, and who can blame them? But it is true.

3

Roses, Perennials and Annuals

The Rosarian Chasm   *17 August 1996*

The rose is our national flower, an icon, a symbol, rich in cultural and historical significance. It is also a garden flower, and probably still our favourite one. We all have an opinion about roses, although not always the same one. In fact, the gardening world is divided into those who like the long-pointed buds, high-centred flowers and recurved petals of Bush Roses, and those who prefer the many-petalled rosette or deep-cup flowers of the Old Roses. It has nothing to do with class, education or refinement, as some people think; it is simply a respectable divergence of aesthetic opinion. No amount of exposure has ever encouraged me to love the flowers of

Hybrid Teas, but I quite believe rose show exhibitors when they tell me, in voices creaky with emotion, that 'Admiral Rodney' is a beautiful bloom.

Those people who like Bush Roses (Hybrid Teas and Floribundas) will sometimes grow a Bourbon rose like 'Madame Isaac Pereire', while Old Rose fanatics may countenance a discreet 'Margaret Merril' or 'Iceberg'. Generally, however, the different races never meet in the garden.

But neither side of the rosarian chasm has ever been entirely satisfied with their lot: growers of HTs would sometimes like richer scent, a laxer, shrubbier habit and less glossy foliage in their roses, to make them better companions for other garden plants, while Old Rose freaks could really do with growing a few which flowered more than once in the season and came in a wider range of colours than just white, pink and purple.

It was David Austin, a farmer in the West Midlands with an interest in plant breeding, who more than thirty years ago saw the advantages of crossing Bush Roses with Old Roses (Gallicas, Albas, Bourbons and so on). The result has been the development of a distinct race of roses, which he calls the English Roses; in the process he has carved the most important milestone in rose history since Guillot introduced 'La France', the first HT, in 1867.

Although it is only twenty-seven years since the arrival of the first proper English Roses (Austin considers his once-flowering 'Constance Spry' of 1961 as a 'stepping-stone'), some are already household names: 'Graham Thomas', which is a delicious rich clear yellow, the strong pink 'Gertrude Jekyll', and shell-pink 'Heritage'.

And there are more to come. Among the best of the newer roses are 'Evelyn' (sheeny apricot-yellow), 'Noble Antony' (deep magenta-crimson), 'Golden Celebration' (deep golden-yellow) and 'St Swithun' (soft pink). Because of the variety of parents used, which now include English Roses themselves, a number of discrete strains have been developed. Although there is, and is intended to be, an unmistakable look to an English Rose, they are not homogeneous in habit or flower shape.

Some roses have been dropped from the Austin catalogue along the way, found wanting because they hung their heads, had too stiff a habit, did badly on certain soils, or succumbed to blackspot. But that is how it should be. It does not make commercial sense for the catalogue to be crowded with superseded varieties.

There are four to six debutant English Roses each year, shown first to the public at the Chelsea Flower Show. Six out of between eighty and a hundred thousand seedlings reared annually from seventy thousand deliberate crosses shows a restraint bordering on self-denial. But gardeners do expect each new one to be different from the others, even if they have sometimes to look closely to see it.

The great strength of the race is that, despite the hundred or so now available, their softness of flower colour, delicacy of flower shape and matt foliage almost guarantee harmonious plantings in the garden. Or that has certainly been the case until the recent introduction of 'Pat Austin', named after David Austin's wife. This flower is a scintillating copper-orange which would look wonderful with purple-blue perennial salvias or rich red-purple cotinus, but the strength and depth of its colour means it is bound to overshadow many of its pastel relations. To be fair, however, we routinely endure far more extreme colour variations in Bush Rose beds.

The development of the English Rose seems to me to elevate Austin to the small, exalted circle of people who have changed the look of gardens: Lemoine by breeding lilacs; Marliac by developing a race of hardy water-lilies; Alan Bloom by breeding a wide range of garden-worthy hardy perennials. Austin is already a prophet with honour in his own country – and abroad. The most popular of his roses have found their way into other rose nursery catalogues. His success lies partly in his timing, for twenty years ago, when gardeners became keen to integrate roses in flower borders rather than keeping them in solitary confinement, he was already producing new roses that would fit the bill. But it also lies in the fact that he has altered an icon to match our heightened expectations. It takes a genius to think of doing that.

Influence of Peace   6 May 1995

Fifty years ago last week, the Pacific Rose Society held an exhibition in Pasadena, California. On the day in 1945 that Berlin fell to the Allies, the Society officially named a new rose 'Peace'. At the same time, in a symbolic gesture bordering on the heavy-handed, two caged doves were set free.

The story of that rose, probably the most famous of all time, is well-known, among older people at least. In October 1936 a seedling flowered in the nursery of François Meilland, a rose-breeder from Lyon. It caught his eye because of its very large, perfectly-formed primrose-yellow, cerise-pink-tinged double flowers and healthy foliage. In 1939 Monsieur Meilland sent some 'eyes' (budding wood) from this seedling to nurserymen in Germany and Italy. Just after war broke out, the American consul in Lyon sent word to Meilland that he was flying home, and had space for a small parcel in the diplomatic bag; Meilland hurried some 'eyes' of his rose seedling to the consul before he left. (This is such an unlikely story that it must be true.) Once in the United States, the 'eyes' were budded and stocks were bulked up and distributed, which is how it came to be named in Pasadena in 1945. When the rose was introduced to the public it was called 'Gloria Dei' in Germany, in Italy 'Gioia', and in France 'Madame A. Meilland'. It is not known for sure what its parentage was, although educated guesses have since been made. It certainly has 'Margaret McGredy' as its pollen parent.

I cannot say that it is my favourite modern Hybrid Tea rose; the combination of yellow and pink in a plant has always seemed to me an unhappy one. Moreover, its scent is faint, and it is inclined to throw blind (that is, non-flowering) shoots.

These are small drawbacks to weigh against its undoubted virtues, however, and I can well understand why it caused such a sensation in the rose world; quite apart from the timing of its introduction, it was bushy and vigorous in habit, and had such large, perfect flowers and disease-resistant leaves (which were then called 'stout-textured' and which we now call 'leathery'). Its fame spread so quickly that it was awarded the Award of Merit by the Royal Horticultural Society as early as 9 September 1947. Its qualities, and its name, ensured massive sales after the war. It was the World Federation of Rose Societies' 'Rose of the Year' in 1976.

'Peace' was not the only plant whose name was influenced by the Second World War, of course. In 1945 and 1946 awards were given by the RHS to a daffodil called 'Alamein', dahlias called 'Arnhem' and 'Battle', a chrysanthemum called 'Gladiator' and, most bizarrely, an orchid called *Cattleya* × 'Stalin' var. 'Victory'. None of these has survived in commerce, though whether it was their names or their characteristics which made them short-lived wonders, it is impossible to say. Fifty years is a surprisingly long time in gardening. Many excellent plants survive only a few years in the trade although they often remain forever, unrecognised, in private gardens, and the vast popularity of the rose as a garden plant and the ease with which it will breed have, paradoxically, encouraged nurserymen to drop any variety which does not sell well.

'Peace' has been such a success because it represented a new and positive improvement in Hybrid Tea roses. Its disease-resistance, especially, attracted gardeners unable to draw on the armoury of effective chemicals we have today. As a result, it is generally credited with the renaissance of interest in rose-growing among gardeners in Britain after the war. Despite an increasing susceptibility to blackspot, its appeal endures. Moreover, it has been used extensively in breeding a better race of garden-worthy Hybrid Tea bushes and climbers, such as 'Prima Ballerina', 'Compassion' and 'Rosemary Harkness'. At exactly the same time as Europe was facing a momentous turning-point, the introduction of 'Peace' marked a significant one in rose-growing. The present younger generation of gardeners may never have known the delights of Clay's fertiliser, Chase's cloches, 'Market King' tomatoes or any other feature of the wartime garden, but they have certainly benefited from a lifetime of 'Peace'.

Pseudonyms' Corner 15 June 1991

It was with regret, but hardly with surprise, that I read recently in *The Times* that Torbay's palms had been rumbled. Emblem of the 'English Riviera' and proud logo of Torbay tourism, they have been found instead to be examples of the New Zealand cabbage plant or *Cordyline australis*. And this, as everyone knows, is in the family Agavaceae, not Palmaceae.

Part of the regret I felt was that Bournemouth, a rival seaside town, should have decided to blow the whistle on what we all knew was a polite but agreeable fiction. I hope that Mr Eddie Hunt, Bournemouth Parks Director (for it was he, apparently, who pointed out Torbay's error), is pleased with a job well done.

Equally, I felt regret that Torquay had put itself in the wrong and opened itself up to criticism when *Trachycarpus fortunei*, which is a true palm, will grow there in any case. This handsome tree may come from China rather than the South Seas, but it is indisputably in the palm family, and suitably palmlike. It has a boss of fronds on top of a bristly ramrod trunk, and would make a fine model for a logo.

I suppose that the problem with the Chusan or fan palm is that it is inconveniently hardy and accommodating; it is possible to grow it in any sheltered spot in Britain, except for the north-east. No doubt Torbay could not countenance the possibility of rival posters appearing in railway stations everywhere: 'Come to sun-drenched Barlaston, where the palm trees sway.'

Even so, it does seem a little hard that Torbay's 'palms' should have been singled out when there are so many other floral misnomers which survive unquestioned. 'Cabbage plant', for example, hardly seems an appropriate name for a monocotyledon with spiky leaves, and 'Jerusalem artichoke' is positively bizarre.

Helianthus tuberosus is not an artichoke at all but a sunflower. What is more, it comes from Canada and the south-eastern United States, regions which arguably comprise the New Jerusalem but hardly fit the description of the old one. The suggested origin of this name seems highly unlikely: that the tubers were introduced from Ter Neusen in Holland in the seventeenth century and 'Artichoke van Ter Neusen' became 'the artichoke from Jerusalem' in the mouths of street vendors. Even the British Tourist Authority could hardly have dreamt that one up. It is far more likely, though also more prosaically, to have come from the French for sunflower: *girasole*.

This outrageous mis-description is just the kind of thing that the European Commission should be stamping out. If it can solve the sticky question of the Cornish pasty to everyone's complete satisfaction, then this should pose it no problem. I look forward to reading Commission Regulation (EEC) No. 12345678/91 of 1 July 1991, which deals with the renaming of 'Jerusalem artichoke' as 'North American tuber-forming vegetable-type comestible'. Such a change should prove popular on market day.

The older the name, the more one has to be tolerant of its eccentricity, it seems. We all accept unblinkingly 'dog's-tooth violet' for *Erythronium*, even though it is not a violet and to compare the root to a dog's tooth is fanciful. The 'mountain ash' is not an ash, nor the 'winter cherry' a cherry, and no one in their right mind would sell fruits of the 'strawberry tree' at Wimbledon. Yet we go along with such nonsense. Although a rose by any other name would smell as sweet, I am glad we have finally abandoned the Middle English 'brier' or 'briar', for these are names also given to *Erica arborea* and even *Smilax rotundifolia*.

There is more excuse, perhaps, for the misnaming of *Pelargonium* as 'geranium'. Two hundred years ago, when first introduced, pelargoniums were thought to be species of the genus *Geranium*. It was not until the explosion of pelargonium breeding in the Victorian age that botanists felt moved to split them off. It does seem perverse for gardeners to persist in this error now, however, when a request for a geranium in a garden centre is as likely to get them a hardy herbaceous perennial as a tender house plant.

The confusions which have arisen reinforce the already strong argument for using Latin names as a matter of course when referring to plants. Torbay's downfall was to pander to the popular clamour for the wider use of vernacular names. If it had stuck to calling its 'palms' *Cordyline australis*, it would not have opened itself up to mockery from Bournemouth – or been troubled by an influx of summer visitors either.

Stagestruck   16 May 1992

Last week I went to the theatre. I enjoyed it enormously, for it was a lively and colourful production, staged before an appreciative, even deferential, audience – of four. At the heart of the action was a mystery: not 'who dunnit', but why an 'auricula theatre' should ever have been built, in 1857, in the grounds of so grand a house as Calke Abbey. For the growing and displaying of auriculas was then the pastime of artisans, not aristocrats.

The many varieties of auricula, which come in every colour

except pure blue, are descended from *Primula auricula* and *P.* ×
pubescens, the scented alpine relations of the cowslip. Their
common name of Bears' Ears derives from their smooth, oval,
inward-curving leaves. The 'show' auriculas prized by enthusiasts have been around since the mid eighteenth century,
thanks mainly to extensive breeding and selecting of *P. auricula*
forms by the handloom silk weavers of Lancashire and
Cheshire. (These were the descendants of the Protestant refu-
gees who fled persecution by the Duke of Alva in the
Netherlands in the 1570s, bringing their auriculas with them.)
Their only recreation was the passionate, meticulous care and
exhibiting of these flowers at local 'feasts'. Such men would
pay as much as two guineas (forty-two shillings) for a fine new
'edged' cultivar, at a time when they might earn no more than
eighteen shillings a week. The family might go short, but
'Grimes' Privateer', 'Popplewell's Conqueror' and 'Lancashire
Hero' would flower each April in the yards of Middleton and
Rochdale.

In the heyday of auricula-growing, the century after 1750,
these flowers were often shown off in 'auricula theatres', which
were usually three-sided roofed enclosures, open at the front,
with shelves in tiers on which ranks of pots could be stood.
Tiny versions which took just two pots were made to be carried
on the grower's back as he walked to shows, held in public
houses.

There is no record of why such a theatre was built into one
corner of the walled garden at Calke Abbey. It is twenty feet
long, eight feet deep and high, and contains eight tiers of
wooden shelves. It originally had blinds, which could be
dropped down in front of the pots in sunny weather – a sensible precaution, as the theatre faces south-east rather than the
more usual north or east.

Some 'theatres' had extravagant scenes painted on the three
back walls, to give interest for the forty-six weeks of the year
when the auriculas were out of flower; the 'scenery' could be

60

hidden by a black cloth at flowering time. At Calke, however, the background is a pleasing, unadorned ochre. This theatre may well have been used for other plants, such as carnations and pelargoniums, to prolong the display, and it is hoped it will be again.

The display in late April and early May is wide-ranging, including border and alpine auriculas as well as the more challenging 'show' or 'stage' auriculas: ' green-edge', 'grey-edge', 'white-edge', 'fancy' and 'self'. The sturdy six-inch stems support heavy heads of flower, which sway in unison in the breeze like gaily-painted chorus girls.

Like actresses, auriculas are a mixture of hypersensitivity and old-boot toughness, from whom the best performance can only be coaxed by cosseting and flattery. Although hardy (as you would expect with a species which originates in the Alps) and quite capable of thriving in the flower border, they spend much of their lives in clay pots under glass, so that no drop of rain can spoil the mealy 'farina' of the silver-green leaves or the sanctity of the 'circle of paste' in the centre of the flowers. Although no longer treated to a rich diet of goose dung steeped in bullock's blood with sugar, baker's scum, night-soil, sand and yellow loam, they are nevertheless still grown in a particular and special compost mixture.

It is a pity that the fashion for auricula 'blooming stages' is long gone, for it seems to me appropriate to display these wonderful plants in theatres with staging and backdrops – not just because they are shown so clearly in all their extraordinary beauty, variety and detail, but because theirs is a story full of artifice and high romance.

Criminal Tendencies   *13 June 1992*

True gardeners are always willing to share a plant with a chum. That is what marks them out from the common, selfish herd and makes gardening so agreeable – or so it is said. Which is all very fine, but I am beginning to think that I have caused more misery with my generosity than ever I could have out of meanness.

As a result of my open-handedness, I have certain friends whom I cannot now look in the face without blushing from unadmitted guilt. One of these is a barrister in Rutland who seeks respite from a gruelling work-load in cultivating his garden. He has had plants from me which, in their own way, show criminal tendencies quite as marked as those of the alleged murderers and bankrobbers on whose behalf he works so assiduously.

Take the yellow hollyhock, for example. I was given a plant of this by a friend, who had it from a famous plant collector, who found it in the Caucasus. (Now, as every keen gardener knows, few things in the world of gardening are more attractive than being able to trace a plant's lineage back to a legal wild collection.) For some years I was happy to give such a distinguished guest garden-room. It had pretty pale-yellow flowers on sturdy five-foot-tall stems in summer, and an enduring bright green corrugated leaf. What is more, it was a model of decorum, setting only as much seed as would ensure its continuing existence. I do not know what happened last summer – perhaps it was the heat – but this plant suddenly began to spill its seed with the abandon of Onan the budgerigar. This year no nook, cranny or crevice is free from seedlings – and plenty of them.

Last year I gave away one or two plants to my friend the barrister. Now, when he goes out into the quiet and calm of his garden to grapple mentally with a tricky jury speech, he must wrestle instead with the tenacious roots of *Alcea rugosa*.

I am not sure he will forgive me this time for passing on an unsuitable plant, for I have as much form for doing so as his clients for assault and battery. After all, it was I who gave him *Phlomis russeliana*, a thug of an herbaceous Jerusalem sage, not to mention *Geranium maculatum,* the tuberous-rooted cranesbill which pops up everywhere. I think I may have to come clean and ask for eleven similar cases to be taken into consideration.

The path to Hell is paved with these attractive but invasive symbols of my good intentions. Frankly, it is unwise to admire anything when going round my garden, if you do not want to be handed a plastic bag containing a generous handful of future trouble.

Even though I acknowledge the problem, I have not yet learned from experience. Only last week, in an attempt to give a expert friend something different, I saddled her with the hairless version of *Alchemilla mollis.* This differs from the ordinary lady's mantle barely perceptibly, but will seed about just as freely.

In mitigation, it has to be said that I have been victim as well as perpetrator of this anti-social crime. The way I acquired the yellow hollyhock and the lady's mantle are examples of that. And at least I have had the sense not to give my lawyer friend *Rumex sanguineus* var. *sanguineus.*

This is the fancy name for a plant known colloquially, and in our household expletively, as the Bloody Dock. When it first appears in spring the leaves are blood-red, and make a good talking-point on tours round the garden. However, by the time it runs up to flower (now), it does not look very different from the dock we used to rub on nettle stings as children, to which it is closely related. This is the moment to cut it back, for woe betide you if it is left to set seed. I was given it by a knowledgeable and generous farmer's wife, who certainly meant me no harm.

Worse still, long ago, before I knew how to decline an offer graciously, a neighbour succeeded in giving me the wickedly spreading yellow-flowered *Sedum acre.* Looking back, I see that was a criminal thing to do. But then, I am hardly in a position to cast the first stonecrop.

Greenery-yallery   *13 February 1988*

Those who have been paying strict attention may wonder why it is that the phrase 'house plant' features so seldom in these essays. That may seem rather surprising. After all, a quarter of all households in this country have more than ten house plants, believe it or not; the most convinced non-gardener will have a rubber plant and adolescent avocado tree in the hall, even if the garden out the back is a wilderness.

The reason I hesitate to write about house plants is that I grow the 'foliage' tropical kinds rather badly and with almost no enthusiasm; nor do I think myself alone in this. House plants do not require enormous skill or even that indefinable quality (so precious to gushy women) known as 'green fingers'. But the prosperity of the sub-tropical and tropical species does depend partly on where you live. All I can offer them is a dark, draughty cottage when most need reasonable light and are as sensitive to draughts as elderly female relations, showing their annoyance even more eloquently than Aunt Agatha, by dropping their leaves.

I have sometimes wondered whether moving house might not be the solution, albeit rather a drastic one, but there are many people who live in light, draughtless, modern houses who have little more success in the long term than I do. This is because foliage house plants also like constant temperatures and high humidity – just the conditions least likely to obtain in houses where the central heating is turned off for more hours than it is on and the occupants have no wish to live in a cloud of steam.

Although house plants do not like houses, the same cannot be said of house plant pests, particularly aphids, whitefly, mealy bug and scale, which come down like the entire Assyrian army the moment a house plant is imported or which, as likely, are brought in with it. Spraying with an insecticide to remove them usually does as much to take the polish off the furniture.

Even if I could give foliage house plants the conditions they want, I wonder how much I would bother. Many find their way into our homes because they will survive (more or less) despite adverse conditions and not because of any marked inherent qualities. In the end, what are *Monstera* or *Ficus* for? Their unrelieved glossy greenness gets one down after a while, the John Innes No. 3 in the pots soon smells rancid if overwatered (which is easily done), and they never flower. Those few which do, like anthuriums, have such unappealing flowers (known in the trade as spathes) that one rather wishes they had not made the effort. I have grown a *Howea* palm for eight years; I feed and water it punctiliously, but I cannot remember when I last thought how nice it was. Am I not growing these plants more for form's sake, or to fit the cache-pots given as wedding presents (which, incidentally, their 'terracotta' plastic containers rarely do), rather than in eager anticipation of continuous pleasure?

House plants come from as great a variety of backgrounds as members of the Labour Party, and they mix no more happily together. The bowls of assorted foliage house plants given as presents last no more than a few weeks before disunity becomes apparent. Quite soon one plant becomes dominant and crowds out its fellows.

The subject of bowls brings me to my greatest objection to foliage house plants: the fact that they encourage the introduction of every kind of worthless tat, from white plastic ridged bowls to macramé plant-holders with tassels.

Of course, it would be quite wrong of me to tar all house plants with the same condemnatory brush. Provided one can find somewhere for them to go when not in flower, many flowering house plants more than earn their keep. *Cyclamen*, *Streptocarpus*, the old cottage favourite *Campanula isophylla*, are all content to live on a windowsill in an unheated passage. Geraniums, or rather zonal pelargoniums, can be shut away in a cold spare room for the winter and scarcely watered and, if cut down to stumps in the spring, will come back invigorated

and refreshed from the experience: being South African mountain or desert plants, the last thing they need is high humidity.
All these can spend the summer outside. I certainly feel that
the shifting about which is necessary is preferable to leaving
great squat lumps of dust-gathering greenery about the house
for years and years.

People like myself who are naturally outdoor gardeners
have turned to dried flowers as an alternative for room
decoration. The 'immortelles' such as *Helichrysum*, *Statice* and
Helipterum, together with ornamental grasses, *Achillea* and
hydrangeas, can be grown in the garden, dried, and arranged
in pretty bowls and baskets. Not only do the hungriest greenfly leave them alone, but they positively dislike high humidity
and, what is more, the darker the room in which they are
placed, the better they retain their colours. So it looks like I
don't need to move house after all.

No Error of Taste   22 July 2000

I first came across petunias in the garden of the British
Residence in Budapest in July, 1973. (Don't ask.) I was very
young and had only recently become interested in flowers and
gardens, but I was charmed by the delicacy and symmetry of
their single, pastel-coloured trumpet flowers, and the generosity with which they flowered. My hostess explained why she
liked them so much: because they could stand the heat of a
Continental summer and would flower for months on end. I,
who had been brought up on old-fashioned roses, 'syringa'
and delphiniums, none of which flowered for more than a few
brief weeks, was intrigued by the idea that a flower might go
on all summer long. (In fact, this was shorthand for a succession of flowers, but I didn't find that out until later, by which
time my attachment to petunias was strong enough to stand
the shock of revelation.) Ever since I acquired my first garden

I have grown them in prodigious quantities, as groundcover fillers in borders and in June-planted pots.

These days it is not uncommon to hear them referred to as 'vulgar', but how could I possibly agree when my earliest memories are of them flowering madly, wholeheartedly, in the garden of Her Majesty's Ambassador to Hungary? No error of taste was committed – that was left to me, and my weakness for wearing that summer (the memory is painful) a brown T-shirt with my childhood nickname emblazoned in red on the chest.

When striving to din some small nugget of illuminating botany into my husband's overcrowded head I describe a plant by reference to its botanical family, saying, for example, that cyclamen is a kind of primrose. Well, a petunia is a kind of potato, although its closest relative is in fact tobacco: 'petun' means tobacco in a South American Indian dialect. As one encyclopaedia primly puts it, 'their leaves have a similar narcotic effect [to tobacco] on humans'. As another of my youthful errors of taste was to smoke roll-ups, it is a good thing I did not know that then, or the diplomatic petunias would not have been safe with me.

The ancestors of our modern garden petunias are three species, *P. axillaris*, *P. violacea* and *P. integrifolia*. They are short-lived perennials with a spreading, straggly habit, up to two feet tall, with sticky-haired leaves and solitary, single trumpet flowers in a limited range of colours, white, pink or violet; *P. axillaris* gives off a sweet, heavy scent at night. They are plainly the ancestors of modern hybrids, but their descendants have larger flowers, with petals often puckered as if made of tissue paper which has been wetted and allowed to dry, in a greater range of colours from white through yellow and crimson to purple, and with either a compact and bushy or a trailing habit.

There have been times in the last fifteen years when my attachment to petunias has been sorely tested, thanks to the perversity of plant breeders. They have concentrated on developing floppy-headed, double, frilly varieties, as well as singles which are either striped or bi-coloured 'picotees', or come in electrifying colours, often with central, darker veining. Not all, by any means, are scented, even at night.These are not vulgar, but they certainly try the devotion of petunia lovers. Moreover, this concentration on developing larger and double flowers means that many are scarcely equipped to deal with damp, cold weather, the petals turning the consistency of wet loo roll and becoming prey to rot.

Nevertheless, many of the newer varieties are definitely worth growing: the single-flowered trailing varieties, such as the famous Surfinias as well as the 'Trailblazer' and 'Wave' series, and the small but generous 'Trailing Million Bells', have an obvious place in pots and other containers. Although I am not wild about those with darker veins in their centres, the delicious scent of the pale blue 'Blue Vein' is a compelling reason for growing it. Unlike a number of other highly-bred 'bedding' annuals, the seed of petunias is available in single colours, so it is possible to devise planting schemes that work, without being lumbered with a weird mix of carmines and orange-reds, as happens with most busy Lizzies. In fact, I also grow Suttons' 'Trailblazer Mixed', because the colours harmonise.

This summer is proving a testing time for petunias, it having been both wet and unseasonably cold. So far this year, the best doers are proving to be, as you might expect, the 'Storm' series (which has now superseded the old 'Resisto' varieties as one of the most resilient of the large 'grandiflora' types), together with the Surfinias and Trailblazers. But I still have to deadhead bedraggled, washed-out flowers regularly, for the look of it. The task is not made lighter by the knowledge that while today it is 17 °C (62 °F) and cloudy where I live, it is sunny and 33 °C (91 °F) in Budapest. Oops – definitely an error of taste to mention that.

4

Garden Design and Gardens

Grit and Determination   *14 May 1994*

One of the more consoling aspects of writing a gardening column is the correspondence I receive from readers. Hidden away as I am for much of the time in an ivory bower, these letters open a window on the wide world of other people's gardening. And not just in Britain, but abroad. In the last ten years I have received several letters from (usually titled) expatriates living along the choicer reaches of the Mediterranean.

The latest letter to arrive is from an Englishwoman in Tuscany whom, as she does not say whether she is Mrs or Miss, I strongly suspect to be a modestly-inclined Lady. She would like advice about making a new gravel bed on ground never gardened before; her concern is how to minimise weeding and watering on what she says, and I can believe it, is very dry land. It is nice of her to ask for my advice. However, as we in Britain have adopted 'gravel gardening' after observing the way plants grow in the wild in southern Europe while on our holidays, I can only hope that no one will think I am taking Sangiovese to Greve.

Gravel sets off plants well and looks clean and fresh, especially after rain. It is as suitable in town as in country, but works particularly well in gardens where stone is already the main building or paving material. The technical reasons why gravel gardening succeeds are these. Because it dries quickly after rain, a surface layer of grit gives the impression that whatever is below it must be dry also, but this is by no means the case. Water soaks in easily enough, but the grit inhibits it from evaporating at the rate it would from bare soil. Therefore, unlikely as it might seem, gravel acts as a sterile mulch, almost as much as bark chippings or cocoa shells, say.

Moreover, as the gravel shields whatever is below from the heat of the sun, a 'cool root run' is assured; Mediterranean plants, used to growing their roots under rocks, appreciate this – as do alpines from countries further north, incidentally. Provided that plants are well watered when they are planted and continue to be so until their roots have become well-established, they should only need to be irrigated thereafter in prolonged spells of dry, hot weather. Yet the necks of the plants remain dry in winter, which ensures the survival of many too rain-sensitive to stick it out in humid conditions in a conventional border.

Plants, both wanted and unwanted, seed very freely in gravel, but weeds are far easier to pull out of grit than out of soil, because of the large size of the air spaces. It is amazing how quickly a gravel bed looks mature.

In a garden I know, where many of the gravel beds are also meant to be informal paths, three inches of unwashed 'hoggin' has been laid and rolled, and then covered with a thin layer of washed pea gravel (a deep layer would be uncomfortable to walk on). Although a crowbar is often necessary to make holes for planting, plants flourish in it.

If the gravel bed is not to be walked on it is not necessary to go to such lengths, although rolling the soil beneath prevents a bumpy surface developing. A one-inch deep layer of quarter-inch pea gravel should be sufficient; a ton will give you that cover spread over about forty-two square yards. Flattish stones of the same kind can be used as stepping stones.

It is wise, though not imperative, to choose plants which are natives of the Mediterranean and accustomed to summer drought and impoverished soil back home. There are plenty of them: all the grey-leaved plants like artemisias and lavenders, together with tulips, irises, cistus, sisyrinchium, euphorbias and Greek peonies.

In Britain, late spring is the ideal time for planting a gravel garden, but that would hardly do in Italy, where autumn, when the rains come, is more suitable. So my correspondent has leisure to marshal her plants, take delivery of a supply of gravel from the local quarry, recruit Mario and Pietro to push the wheelbarrows, and, if necessary, borrow a crowbar, before it is time to plant. I am prepared to bet that, were she to invite her friends and neighbours round to admire her gravel bed this time next year, even those of rank in Tuscany could scarce forbear to cheer.

Adventure Playground   *11 March 1995*

Ask a garden designer in what direction she (for they are often she) thinks garden design is going in this country, and the chances are that she will reply 'Backwards'. Whatever her own preferences, her clients will be asking for a 'cottage' garden here, a 'formal' garden there, an herbaceous border, even, somewhere else. So-called 'contemporary' garden style is for public and corporate spaces, if at all.

According to David Stevens, the Professor of Garden Design at Middlesex University, the backward-looking, essentially plant-oriented approach to garden design, which is so often what the client seems to want, is holding designers back. He is quite right, of course. Recently, in the trade magazine *Horticulture Week*, he deplored the lack of innovation in Britain, saying that garden design, as practised by the professionals, was seventy-five to a hundred years behind the times. (He was referring, no doubt, to the still powerful influence of Miss Jekyll, with her colour-themed borders and emphasis on profusion and informality within a formal setting.) He believes that part of the reason for this is that the Modern Movement has had far less impact in this country than in the United States, whence he draws much of his inspiration.

Gardens in Britain are too often, therefore, pastiches of earlier styles, wildly inappropriate to modern life especially because they are maintenance-intensive and reliant on expensive 'traditional' materials of limited availability. He would like to see much more use of newer materials – plastic for flooring, polyester for fencing, even lightweight alloys for buildings which can be covered with rot-proof translucent fabrics – and the more imaginative use of theatrical garden lighting.

David Stevens is an engaging and energetic chap with whom I have happily collaborated on a book and, unlike so many of us, he is not unduly burdened by cultural baggage. He sees gardens as playgrounds and 'outdoor rooms', not refuges from the unfriendly world. That is why he has little time for cottage gardens, with all their complex resonances. If I understand him correctly, plants should be used for structural effect, or as contrasts to the 'hard' features; their individual beauty can be secondary. The best kind of garden is a series of connected spaces, contained and softened by planting but never dominated by it.

There are some surprising similarities between the Modernist approach and that of the 'formalists' who take their inspiration from seventeenth-century gardens (although I should be surprised if they would thank me for the comparison). In the formal, symmetrical, highly structured garden, the individual identity of a plant is often subsumed – into a hedge, into an avenue. Simple geometry is the keystone. The difference is that, in the formal garden, the shapes are mainly linear or circular and are underpinned by traditional 'hard' materials – gravel, stone paving, handmade bricks – while in the contemporary garden shapes are often curvilinear, spaces are asymmetric, concrete and timber predominate, and plant groupings can be fluid 'rivers' of colour, *à la* Burle Marx. In the 'contemporary' garden stress is often laid on its place in the wider landscape, and native plants are positively encouraged. More so than in the United States, where the native flora is colourful, the use of wild flowers and native trees in Britain lends the contemporary garden a more muted aspect than one planted on Jekyllian lines.

Why are we so resistant to a more forward-looking approach? It must be because most of us feel we only have one or two stabs at creating a garden in a lifetime; we do not want to make an expensive, time-consuming bish, so we fall back on a style of gardening that we have seen works in our climate. We open our gardens to each other, so the same ideas bounce back and forth. Intellectually I appreciate David Stevens's position, and do usually find good examples of 'contemporary' gardens stimulating, but I have not the nerve to make one myself.

In Britain we have a tradition of spending more time than money on our gardens. When a progressive designer tells us it will cost £15,000 or more to lay out a garden, what with the timber decking, the pool and the polished granite bubble fountain, we practically pass out at the very idea. In fact, our traditional formal garden, with its hedges and trellis, though cheaper to make, may well end up costing more to maintain.

Most alarming to us, the 'contemporary' garden both requires a great deal of forethought and, once made, it is made. The great advantage of a plant-dominated 'cottage' garden is that it can be fiddled around with, endlessly. It is never finished, and that suits most of us just fine.

Thyme to Bow Out   *15 April 1995*

I like to see myself as a regular kind of a girl, mainstream and uncontroversial in my opinions and amiable in personality. I am the sort who can listen to somebody's barmy idea about her garden without feeling the need to tell her how barmy, who is capable of receiving the kind offer of a *Dieffenbachia* gracefully, and who is always happy to listen sympathetically to those trying to grow courgettes in heavy clay or artemisia in the shade. Why is it, therefore, that when faced with the seemingly innocuous subject of the herb garden I feel so out of joint with the world? So much so that I don't care whom I offend by saying so.

Herb gardens are in fashion. There are more in Britain now than there were in the Middle Ages, when we needed them. People derive tremendous fun, even moral uplift, from creating and tending them, and then doing something constructive – or simply something – in due season with the horehound or elecampane they have grown.

I don't mind the fun; it is the moral uplift which makes me grumpy. For just occasionally, in someone's herb garden, I have sniffed the odour not of agrimony but of sanctimony. I know perfectly well (before any of my herb-growing friends write in) that there are plenty of delightful, open-minded, amusing people who grow herbs seriously, but they undoubtedly attract the self-righteous and the numbingly earnest, too. Herbs, and the making of gardens to put them in, too often go hand-in-hand with caring, non-violent (of course!) concern for animal rights, and non-exploitative shower gel.

Although no one could deny that we need a number of culinary herbs, and can benefit from several with medicinal or cosmetic properties, I do wonder why there has to be such a song and dance about growing them. Why do they have to be dignified with their own garden, at the best paved and hedged, at the worst planted within the spokes of a horizontal wagon wheel? Herbs look pretty in the spring with their fresh newly-minted leaves (then again, so does the rest of the hardy plant world, with the possible exception of the yellow-leaved form of *Sambucus*), but by high summer they are as dusty and seedy as a Soho night spot. Surely they are best put in mixed borders where they can play strong supporting roles in spring and early summer, but withdraw discreetly to the back of the stage to murmur 'rhubarb, rhubarb' to each other for the rest of the season. Where there are no suitable borders, the kitchen garden, the window box or pots on the patio should do.

If you decide to make a herb garden you will feel you have to include most of what are considered herbs, but some of these are a positive menace. Rue causes severe skin allergies in some people, borage is enchanting but should rule itself out because it seeds everywhere, horseradish is coarse-leafed and as ineradicable, once established, as bindweed, and the variegated form of lemon balm looks as if weedkiller has been inexpertly applied to its leaves. There are few, a dozen types at the most of common culinary herbs, which are pretty enough to deserve a place for their looks alone. The question I think people should ask themselves, therefore, when they make a herb garden, is whether they wish to develop an outdoor medicine cabinet, larder, historical laboratory, or attractive garden. The first three may rule out the last. If you are not to throw all ideas of aesthetics out of the window, you must liberally interpret the word 'herb'.

That means including anything that someone else has found

useful, such as *Hemerocallis* (daylilies, which the Japanese fry and eat), Australian eucalypts, *Rosa damascena* 'Trigintipetala' (Attar of roses), peonies (which are used in Chinese medicine) and those attractive border plants *Alchemilla mollis*, which was once used as a cure for toothache, and *Anthemis tinctoria*, from which a dye can be extracted. It also means abandoning the purist approach sufficiently to discourage those native herbs which appear of their own accord and spread like crazy, such as coltsfoot, shepherd's purse and black nightshade.

The trouble is that if you take a liberal and aesthetic view of what a herb is and what a herb garden may contain, you might just as well call it a 'summer garden' instead. If you do that, the real thing can be safely left in the hands of those worthies who are keener to dye their own wool with a hand-prepared mordant derived from horsetail than to create a pretty garden.

Don't Think Singles   *25 May 1996*

No plant is an island, entire of itself. In the wild, plants grow in groups or colonies, sometimes so numerous that their communities stretch for miles. Where they do occur singly it is because of some hiccup, such as loss of habitat, attack by animals, or voracious collecting. The vast majority of wild plants require, or at the very least benefit from, cross-fertilisation with others of their species, in order to mix their genes. If you see a plant on its own, you can bet that its future is uncertain.

In gardens, things are very different. We may pay lip-service to the idea of naturalism, but the modern garden is often no more naturalistic than a breakfast television studio. In most, the singleton rules, OK. One philadelphus, one deutzia, one aucuba, one forsythia: these are the common inhabitants of the shrub border. Of course, lack of space will usually prevent a truly naturalistic approach, but we could perhaps accept, more readily than we do, that a group of identical plants is better at making a strong visual statement than a hotchpotch of individual species.

Even large trees (which are the plants most advocated by experts as suitable for 'specimen', i.e., single, planting) would often look better as part of a small grove instead. We would all, I am sure, rather see a grassy orchard than a single apple tree. Many a large lawn could be improved if the single specimen tree which punctuates it were joined by a couple of fellows. Only the mulberry and other pendulous-branched trees, whose shapes are so singular that they could not easily accommodate companions, are exceptions.

In 'mixed borders', where herbaceous perennials, rock plants and bulbs predominate, there is far less excuse for single, pimply planting. Yet it is widely practised, especially in so-called plantsmen's gardens. Here 'difficult' plants are prized, but these often palely loiter, without the strength to reproduce or make a decent clump quickly. A real plantsman's border is as bitty as a chocolate praline.

Even if we consider we grow plants as much for their overall effect as for their individual interest, opportunities are still missed. We unhesitatingly plant herbaceous perennials in uneven-numbered clumps, because we have so often been told to do so. That is a helpful suggestion (no more) with slow-creeping perennials, but with vigorous ones it is unduly restricting. In fact, it does not usually matter whether a clump of asters consists of three, four, five or sixteen plants; once it is established, you will find it hard to identify each individual anyway. And a group of six geraniums, for example, is no less likely to make a fluid pool of flowers than a clump of five. But if all your clumps are in threes or fives you may notice a deadening sameness about the border. In nature there is a constant ebb and flow, caused by competition between species and aided by climatic variations. We could sometimes try harder to emulate that.

With strict adherence to the uneven number rule often comes a reluctance to replicate groupings. If I plant three delphiniums in a group I think that I have done my duty, when really I should be planting several clumps of three delphiniums, or one clump of three and one of six or seven. Nothing visually unifies a border better than the repetition of a few successful 'incidents'. Without repetition, the eye has to work harder to make sense of what it sees. If garden space is so limited that the repeating of groups seems excessively constraining, something to the same effect can be achieved if plants of similar foliage shapes or flowers are put together.

Repetition is not really such a strange idea, even in the context of a highly artificial garden. You only have to think of hedges, which provide so much important structure in the garden; to achieve the required symmetry, these must almost always be composed of a single species of plant. Or better still, think of the lawn: a collection of identical individual plants repeated almost *ad infinitum*. If gardens are to be places of beauty and repose, rather than simply poor imitations of botanic gardens, we need to accept the notion that less usually means more.

Style Victims   25 January 1997

What do you think is the dominant garden style in rural Britain these days? Sophisticated 'cottage garden', mock-Tudor knot garden, informal woodland? If you live in the country, you will know that it is immaculate green sward with scattered 'island beds' and specimen conifers.

Gardens in this style can be seen in every village, hamlet and sold-off field in the country. They are all too easy to spot. They can be attractive, and are often unexceptionable, but they have basic shortcomings. At their worst they are divided from countryside and farmland by a line of false or Leyland cypresses, rather than a hedgerow of native shrubs or trees; the flower beds are colourful jumbles of disparate elements, with plants separated from each other by expanses of brown soil; there is often an irregularly-shaped pond marooned in the middle of the lawn, its water plants contrasting oddly with the standard roses in the bed next to it; fruit trees are either stunted, as they are grown on 'dwarfing' rootstocks (as if the garden were a commercial orchard), or they are of the ramrod, unbranching 'Ballerina' type.

The imperatives of soil and aspect are ignored: much trouble is taken to create peat pockets on limestone so that some rhododendron can struggle, while modern ideas of 'ecological' planting (groups of the same plants growing in dense communities, on soil that really suits them) do not feature. Many of the plants seem chosen more for their unvarying nature, what the gardening magazines call 'all-the-year-round appeal', than their inherent attractiveness, thus ensuring that the differences between seasons are blurred.The lack of continuous shelter and reliable food sources makes these gardens unattractive to wildlife.

Their gravest shortcoming, however, is that they make no concessions to locality, or even their immediate setting. There seems little appreciation of the need for a softer, more sensitive approach where village or house meets country. Even where the style of vernacular buildings is very particular and local, gardens look the same as elsewhere – namely, exotic.

The silly thing is that these gardens can also be rather hard work to maintain, for they only look impressive if they are frequently attended to. Sparsely planted, and laid out so that they can be seen at a glance, they depend for their impact on a well-mown lawn, sharply edged, well-behaved trees and shrubs, and neat summer annuals. Anarchy extends only as far as the mixture of colours used. A browning conifer or a variegated shrub whose leaves have reverted to green will catch the eye disconcertingly. The appeal of this kind of garden for its maker must lie in the fact that it does not have to be 'designed' but develops haphazardly, and that its successful maintenance, though hard work and expensive because of the need for power tools and chemicals, is not problematic.

This style emanates from, and has its place in, the town or suburban garden, where its verdancy is attractive, and where its lack of naturalness doesn't strike nearly such a false note. But it has been carried into the country by people who have moved out of towns, just as the seed of the Oxford ragwort was carried round the country in the slipstream of steam trains, carelessly but inexorably. This kind of garden is even promoted by some garden writers and designers who do not wish to appear snobbish or exclusive, and has certainly been helped by the spread of garden centres.

Yet, although the incongruity of such gardens in the country can strike the onlooker as forcibly as uPVC patio doors in a Georgian façade, no planning officer rushes in and cries 'Don't do it!' Indeed, garden owners often have before them the example of parish councils, who are inclined to fill any spare green space in a village with purple-leaved cherry trees and highly-bred trumpet daffodils.

Most people have a sense of place. Anyone who has lived for any time in Northamptonshire (to take the county where I live as an example) can easily divine the reasons why Litchborough does not resemble Fotheringhay. Yet, though we have no difficulty discerning the differences between limestones, especially once they are pointed out to us, it seems we still cannot distinguish between town and country. And things can only get worse.

Moving out of the town should entail a recognition that some ingrained gardening habits are inappropriate. Just as it is necessary to adjust to a life without opera and pizza deliveries, so we must learn to live without standard roses and yellow-leafed conifers. Otherwise, we kill the thing we love. It is as simple as that.

The Importance of Planning *19 September 1998*

Gardening pundits grow prematurely old and querulous impressing on readers and viewers the importance of planning what they want in their gardens, before they go out and buy the plants to fill them. Nevertheless, people can still be overheard in the garden centre, deciding whether the evergreen euonymus or the Russian vine, which look about the same size in their pots, would be best for climbing up the front of the house. If there has been no prior research, gardeners usually have only the plant's label to tell them its botanical name, its likely height and spread, the season and appearance of its flowers, the soil and aspect which suit it, and its initial and subsequent cultivation needs. On the usually small label (which if the plant is selling in the EU will be in four languages anyway) there will be no room to expand on whether it displays other attributes than the flowers or fruits depicted in the tiny photograph, or even whether it has shortcomings, such as too-vigorous growth or over-enthusiastic seeding. Unless the buyer knows the plant well already, the opportunities for later disappointment are legion.

In the hope of improving matters, a process of research and consultation is presently being pursued by *Gardening Which?* (the gardening arm of the Consumers' Association), with the co-operation of the Horticultural Trades Association, which represents garden centres and nurseries. Surveys, focus groups, interviews, and field trials have all been brought into play, and designers, plant experts, semioticians (crikey) and editors are also involved, in an attempt to develop a more helpful, accurate, and even truthful label.

Commendable and worthy as this campaign undoubtedly is, I suspect the complete answer will prove elusive. No label, however well-considered, can possibly address the unique nature of each garden, nor our generally profound ignorance on the subject of our own.

Think of soil, for instance. A row of small gardens, even if laid out at the same time, may not be identical; old-established gardens are often highly individual. I have at least three distinct types of soil, and there may be even more buried under the lawn which I don't know about. Soil is a very strong determinant, yet most of us have no idea of the exact pH, nutrient profile, humus content, and water-holding capacity of our soil or, more likely, soils. Often it does not matter, mercifully, for many plants are forgiving; but sometimes it makes the difference between life and death.

Then there is 'aspect', by which we mean the direction in which the garden, or part of it, faces. For instance, we make the mistake of thinking that just because a garden lies to the south of the house it is necessarily sunny. Exposure, or lack of it, to wind and air currents also matters a great deal, as does rainfall (both its seasonal incidence and its amount), snow or the lack of it, the minimum mean winter temperature, how easy or difficult it is for frost to drain away, and the likelihood of late frosts. This is a complex and individual mix of factors whose interactive influence on the future of the plants we blithely place in the garden-centre trolley is immense.

I sometimes take less account of these factors than I should, and I have no excuse. To quote just one gloomy example: I was lured by a vague but upbeat label into buying a frost-hardy perennial, *Strobilanthes atropurpureus*, in a nursery. I had heard the name, and the fact that it was 'choice'; that was all. I planted it, as the label advised, in a relatively sheltered (for my windy garden) place, in sun and a lightish soil. It comes into leaf late in the spring, grows two to three feet tall and has sparsely-borne purple salvia-like flowers in early autumn. I was perfectly happy with that, as I knew no better, until one day in August I saw a specimen in Devon which was five feet high and covered in sumptuous flower. If only the label had said, truthfully, 'You must be barmy if you think this plant will flourish in a lightish soil in a cold, exposed garden like yours in the East Midlands'.

Plants are not so much curtain material, which may be

chosen from a swatch and whose colour, texture and longevity are assured. With such various origins, and subject to all the vagaries of living things, it is not surprising if they sometimes react badly when dumped down in conditions which differ even slightly from those in their native habitat. That being the case, it won't be a new type of label, even were it the size of an estate agent's sale board, which will save me from further disappointments, unless I can combine its information with a sharp eye, an inquiring mind, a humble spirit and, yes, meticulous research and planning.

What I Did in the Holidays   19 September 1992

If there is one thing worse than going on holiday, it is coming home again. No gardener likes to leave behind the maturing fruits of so many months of effort, still less to return to stringy beans and towering sow-thistles.

This year, however, things were not as bad as usual. The rain which fell on the west coast of Scotland fell too on the dry East Midlands, so that, although the beans were stringy and the weeds dreadful, I returned to a more floriferous garden than I could confidently have expected. It was an instructive lesson in the way vain hope can sometimes triumph over compelling experience – but not half so instructive as visiting woodland gardens in Argyll at the 'wrong' season.

There are no other gardens in the British Isles thought to be so exclusively 'spring' gardens as those to be found on the west coast of Scotland and, in particular, Argyll. The almost frost-free climate, the 90 inches of rain annually and the acid soil combine to suit pre-eminently the Asiatic rhododendrons; and most rhodies flower between April and June. (There are a few exceptions: *R. dauricum* and its offspring *praecox* can be out in January, while 'Polar Bear' swelters in August; but the general point remains true.)

There are almost a score of these gardens, and even their names have a potent romance for a holiday Scot like me: Arduaine, Ardanaiseig, Achamore, An Cala, Ardmaddy, Achnacloich, Ardchattan, Barguillean, Crarae, Strone. One year in May I visited some of them, and was forcibly struck by their rugged and foreign grandeur as I wandered in 'Himalayan' glens among huge tree rhododendrons so covered in flowers that you could not get a penknife blade between them. I came home with traveller's tales of leaves eighteen inches long (*R. sino-grande*) and trusses of flowers the size and shape of a football (*R. macabeanum*).

I did not expect a repeat of such excitement during an averagely wet and windy Argyll August. Certainly I visited some gardens, but more out of respect and force of habit. I had a mild curiosity to see those traditionally Scots walled enclosures where vegetables, fruit, cut flowers and a windswept herbaceous border or two might compensate for the lack of springtime floral fireworks. In the event, these were interesting – especially for the variety and quality of the cabbages grown – but it was plain that the climate is unkind to domestic gardening here.

The big surprise was how impressive still were the 'wild' parts of these gardens, even when largely out of flower. This is the legacy of close care taken by long-dead Glasgow magnates and Campbell lairds over layout and construction, and the full use they made of natural features and settings. However substantial the shelter-belt, there is always in these gardens somewhere where you can catch a glimpse of Loch Etive, Loch Melfort or the island-studded Firth of Lorn.

Even keen gardeners are too ready to be charmed principally by flowers, when we know quite well that these are only the thin icing on a very rich cake. In spring, I might well have missed the fox-red indumentum (hairs on the undersides of the leaves) of *R. eximium*, or the cinnamon bark of *R. thomsonii* And in spring I should hardly have glanced at the handsome and lofty trees which provide the shade and protection so necessary to rhododendrons.

In any event, it would be misleading to say that there was nothing 'out'. August has its own, albeit limited, range of flowering shrubs, some of which thrive in a warm wet climate. Hydrangeas of every kind look far better here than in dusty front gardens in southern towns: the hortensia flowers turn a bright litmus blue when grown in acid soil. More than once on a garden tour I startled my family with a cry of 'Eucryphia!' and they were left to wonder why they should be cryphing, and where. If a tall, stately, white-flowered shrub-tree like *Eucryphia glutinosa* could be persuaded to grow in any soil or climate, it would be as populous in the south as laburnum and a great deal more welcome.

Gardens on the west coast of Scotland which open to visitors usually donate at least part of their takings to Scotland's Gardens Scheme, so information on them may be found in *Scotland's Gardens*, the Scheme's Yellow Book, published annually. Most get far fewer visitors than comparable gardens in the south. So, not only can you come across unfamiliar plants growing with great vigour and dash in a beautiful setting, but you often have the place to yourself – whatever the time of year.

The following was prompted by the death of Nancy Lancaster, at the age of ninety-seven.

Midsummer Dream   17 September 1994

I knew nothing about interior design. If asked, I would probably have said that Colefax and Fowler were a comedy act. Twenty years ago, all I knew about Nancy Lancaster, who owned the firm after the last war, was that she was prepared to pay me fifty pence an hour, which was generous considering my lack of experience, to work in her garden. It was late June, my twenty-first birthday, and I was happy.

I remember a great deal about that first day: I picked goose-berries, tied in raspberry canes, painted the Latin names of herbs in white on wooden labels, weeded a border full of milk-weed, which made my eyes very sore when I foolishly rubbed them, and cut a finger almost to the bone with my new pruning knife. After years of Modern History here was proper practical work, but with respectable intellectual and aesthetic proper-ties. I had stumbled, swiftly and painlessly, on the reason why most people turn to gardening.

I worked with Tom Chalk, a local Oxfordshire man in his seventies, and the head gardener, Mr Clayton, whose first name I never knew, who was a Geordie in his late fifties. They wore an identical uniform of waistcoat, collarless shirt, trou-sers with a belt, and flat cap. They were gruffly kind to me. Mr Clayton had been bred in the tradition of private service gar-dening and saw it as his duty to teach me as much as he could in the few months I was to be there. I worked harder to please those two than ever I had for my tutors at university.

By the time I met Mrs Lancaster she had been some years in the renovated Coach House, having moved out of Haseley Court, the eighteenth-century mansion from which she had famously flown the Confederate flag. She still, however, exer-cised some influence over the garden. That first day, Tom Chalk took me 'next door' to see the remarkable Victorian topiary of chessmen set out in the middle of a game. Even I, who knew so little, could see that I had landed somewhere unique.

The Coach House garden was the outdoor equivalent of the 'English country house interior look' for which Mrs Lancaster was, apparently, famous, despite being an American. Partly walled, it was laid out formally in four segments, with painted wooden trellis pyramids for climbing plants and central, open, blue-grey wooden summer-house. The beds were edged with box and filled with old-fashioned roses, herbaceous perennials and interesting tender annuals, like the lime-green *Nicotiana*: all was airy, light, and harmonious, even though strong colours were not disdained. I remember particularly a *Rosa mundi* hedge, bordered with pink, crimson and white sweet Williams, and *Rosa filipes* 'Kiftsgate' (called 'KitKat' by Mr Clayton) as a forty-foot foaming cascade of flower descending from an old yew tree.

Another marvel was the laburnum tunnel. I had not then heard of Bodnant, and thought it a miracle of imaginative planting. Mrs Lancaster had set a mirror into the wall at one end of it, to appear to lengthen the tunnel. This hardly added to the fun of pushing a wheelbarrow along it, for it seemed to go on forever.

Despite the beauty and congruity of the garden, or perhaps rather as a result of its complexity, I loved most the fascinating mix of tasks which were part of any day's work: scaling lily bulbs, pruning pyramid apple trees, planting freesia corms. Memories of all that gave me courage in the days of aching boredom spent cutting lawn edges at Kew.

It must have sometimes been wet that summer, since I learnt the expression 'Aylesbury rain' from Tom for a day when it never stops. And I remember being soaked by the box-hedge edging when reaching over to weed. It was no hardship when it poured, though, because there was always some intriguing little task of mystifying purpose to learn in the display green-house, which smelt of John Innes No. 3, dying leaves, raffia, and nicotine insecticidal smokes.

Mrs Lancaster, then aged seventy-seven, was often in the garden; she seemed a most elegant creature to me, with her wide-brimmed hat, aquiline nose and very keen look. Even I, with my most unworldly upbringing, could see she had style. The stories about her were legion – mainly concerning the hospitality at Ditchley Park before the war ('the taps were gold, you know'), her generosity and her lack of snobbery. I discovered soon enough that she was a Virginian, a niece of Nancy

Astor, and a great anglophile. (There is a parallel here with Lawrence Johnston at Hidcote, another American who made a memorable garden to suit the English climate.)

Once, remarkably, she invited me to lunch, where she told a succession of witty and slightly wicked stories, none of them about gardening. Despite my blindness to interior decoration, I do remember the Coach House as being full of flowers and very comfortable. One very hot day, as I mowed the lawn near the house, the butler brought me out a glass of sangria on a silver salver.

I have never been back. Not because it was not possible, but because it might have broken the spell of enchantment which surrounds that summer for me. I hated the thought of arriving as a visitor on an Open Day, only to discover Tom Chalk and Mr Clayton long since retired and Mrs Lancaster old and perhaps frail. And the garden, so often imitated, might no longer seem so original and so perfect.

Stately Pleasure Domes

21 October 2000

Domes, however stately, don't always get a good press. And, by association, nor does any organisation caught up with them. But for me the reputation of the Millennium Commission remains untarnished, for it had the sense to give around £40 million – admittedly at the second time of asking – to help build a series of connecting domes in Cornwall.

Only those who have difficulty putting a name to the Prime Minister will not have heard of the Eden Project, under construction in a worked-out china-clay pit near St Austell. Having seen the work in progress, it is certainly hard now for me to relive the astonishment I felt on first reading of Tim Smit's plans in 1995. (Tim Smit, I am sure I don't need to tell you, is the Tigger-like enthusiast who found the 'Lost Gardens of Heligan' ten years ago, and then went on to think up the

Eden Project.) His idea of creating a large number of artificial, climate-defined zones representing Amazonian tropical rain forest, Kenyan savannah, Mediterranean olive grove and so on, in the world's largest greenhouse complex, in order to teach people the relationship between themselves and plants and the importance of sustainability, seemed to me brilliant, but almost certainly impractically ambitious and likely to founder on lack of money.

I was wrong on both counts. In 1994 Smit, a publicist of genius, who thinks and talks huge, teamed up with a West Country architect, Jonathan Bell, and persuaded Nicholas Grimshaw and Partners (designer of the Waterloo International terminal) to tender for the job. The engineers Anthony Hunt Associates and Ove Arup, the landscape architects Land Use Consultants (who recently worked on the Princess of Wales Memorial Garden) and the builders McAlpine Joint Venture (Sir Robert with Alfred) joined in. First-rank botanists and gardeners such as Professor Ghillean Prance, Philip McMillan Browse and Peter Thoday were called in to advise. As for the money, the £80 million capital project has been funded by a mixture of National Lottery, European, and commercial sponsorship money, together with donations and loans.

Grimshaw came up with the idea of two enormous geodesic domes linked by smaller ones, the one mimicking the atmosphere in the humid tropics, the other the warm temperate zones. These are called covered 'biomes', a term usually reserved for an eco-system like a tundra or steppe. From a distance these conservatories look like giant bubbles, captured in chicken-wire. The third 'biome' is uncovered, and is a series of crescent-shaped terraces for showing off the flora of Cornwall, together with that of other temperate regions. Bodelva Pit is a fifteen-hectare, fifty-metre-deep south-facing bowl plainly suited, once filled with soil, to growing a range of tenderish plants. There is also a two-thousand-seater amphitheatre for open-air concerts and events.

The Eden Project is a feat of design and engineering which out-Domes the Dome, and at a tenth of the price: the steel biomes, one of which is 50 metres tall and 110 metres wide at one point, have no internal support and are 'glazed' with hexagonal panes made of layers of ETFE (ethyltetraflouroethylene), a transparent foil which is lighter and safer than glass and has a life span of twenty-five years. Because it is anti-static it is self-cleaning, apparently. The bowl is below water-table level so the drainage system has to be sophisticated; the ground water is filtered through matting so that it can be recycled. Nor must the gardeners be forgotten: many of the plants, including a fifteen-metre-high kapok tree, have been grown at the Project's nursery nearby. Early October saw the first plantings in the tropical 'biome'.

The Visitor Centre was built early on, so that people could see 'The Big Build' before the site opens fully at Easter 2001. I took my family there this summer. They were captivated, which is no easy feat where matters horticultural are concerned. We even went along with donning fluorescent jackets and (completely pointless) hard hats, sitting in a kind of open train, listening to a tape of Tim Smit telling us what we were seeing. We never got very near the site, but we felt somehow involved. We watched as the 'sky monkeys' put hexagons in place, then went to the Visitor Centre to listen to more right-on right thinking and look at a number of clever and amusing interpretative exhibitions.

Apart from the balmy climate and its proximity to Heligan, the great advantage of Bodelva Pit is that it is in a 5B zone, entitled to lashings of European Regional Development Fund money. It needs it. This is not the relatively prosperous, tourist-oriented Cornwall of Padstow and Polruan. Except that the slag heaps are white, not black, the nearby pit villages could be in South Yorkshire. It is interesting that the project got under way when Restormel Borough Council gave a £25,000 donation, putting its faith in the Eden Project as employer and visitor draw, predicted to attract 750,000 people a year.

This is a venture committed, therefore, to industrial regeneration, international scientific research, sustainability, conservation, education, arts and entertainment. All on a very big scale. If it does not collapse under the weight of its diverse objectives, it might just become – in Tim Smit's words – the Eighth Wonder of the World. And Wonders of the World are what Tiggers like best.

Some Like it Hot

It is a curious feature of gardeners that we should be so fascinated by plants which grow in a warmer climate than our own. On the domestic front, this means cultivating house plants (mostly badly, because we cannot provide the right conditions) and, when we visit large public gardens, heading in a straight line for any display greenhouse or conservatory which is dedicated to the care of sub-tropical and tropical plants. This is, if anything, even more true of North Americans and continental Europeans than it is of Britons.

Admittedly, there are far fewer 'stove' houses to visit than there were in Victorian times, for the ruinous cost of heating has done for any which were not cocooned in botanic gardens. But there are plenty of temperate conservatories about, where a sufficiently high temperature is maintained to ensure that frost-tender plants from everywhere bar the equatorial belt will feel at home.

Some of these have to be sought out. The conservatory at the top of the Barbican Centre, for example, is a remarkable and generously proportioned structure, almost an acre in size and stuffed full of good plants. Yet it must be one of the best-kept horticultural secrets in London. Perched above the Guildhall School of Music and Drama and the Barbican Theatre, and sparsely signposted, it seems to escape the notice of most people who come to the complex. This collection of intriguing and ornamental plants is visited by only two or three hundred people a weekend.

Opened in 1984, it was designed by Chamberlain, Powell and Bonn, the architects for the Barbican Centre, and its steel structure made by Hawker Siddeley. The lush vegetation it contains effectively disguises the many services – heating pipes, fire escapes and so on. Skilfully have the ugly liver-purple engineering bricks and white structural supports been masked or covered by creeping, twining or cascading vegetation, such as orchids, bougainvillaea, tradescantia and the Swiss cheese plant (*Monstera deliciosa*). This is the place to see what your house plants really could do if set free from the constraints of poor light, a dry atmosphere, cold draughts and small pots. The conservatory is sixty-five feet from floor to glass roof at its tallest, which allows for the almost untrammelled growth of avocado trees, grevilleas, bananas, banyans, palms and the lovely but tender grey-blue Kashmir cypress.

The superintendent of parks and gardens for the Corporation of London, part of whose job is to oversee the conservatory, is a soft-spoken Welshman called David Jones, with a liking for Shakespeare and unusual plants. He swaps the latter with such scientific institutions as Kew Gardens, and has attempted, with informative plant labels, to educate what visitors there are about the origins of the plants. It is his idea to house the National Collection of fuchsia species here.

A minimum night-time winter temperature of 50 °F is maintained to discourage the tropical plants from dropping their leaves. Both watering and 'misting' (water-sprinkling to increase humidity) are automated, and nitrates are added, when needed, to counteract the alarmingly alkaline nature of London water. Other fertilisers are also fed to the plants in this way. The care of such a mixed collection of plants, partly grown in beds and partly in pots, makes a flexible regime necessary. Most pest control is done with natural predators and parasites, including some ferocious terrapins which eat the cockroaches.

Cacti and epiphyllums are kept in the Arid House, on the second level. There used to be fuchsia varieties here, but water would occasionally fall on the stage of the Guildhall School beneath, so drought-lovers seemed preferable. The most celebrated member of the collection is a seventy-year-old Saguaro (that tall, branched cactus seen in all the best Westerns), a gift to the Queen Mother from the mayor of Salt Lake City.

In such a large conservatory as this, it is possible to display 'house plants' in all their innate rude vigour, scrambling, trailing and hanging in veils – not, as we usually see them, etiolated and scorched on sitting-room window sills. Recently a climbing doxantha (bignonia) managed to find its way out through the vents in the glass roof, so perhaps one day an aristolochia or a passion flower will trail down the stairs and startle the theatre-goers. It might encourage them to take a closer look.

Oh, to be in Holland   *25 March 1989*

For many years, one of the more popular destinations for British coach parties in the spring has been the bulb fields of Holland or, as the Dutch (whose English is really only distinguished from ours by a better grasp of our grammar) would say, 'the bulub fields'. About a quarter of a million of us visit them each year, and it is easy to see why. In January the coastal strip between Haarlem and Leiden is washed clean of colour, a flat landscape in monochrome relieved only by the terracotta red of pantile roofs. But from early March, when the crocuses first begin to flower, this landscape is transformed into rectangles of bright, distinct colours, like a giant-child's patchwork blanket, bordered by the grey sand in which the bulbs grow so well.

The British are divided into two camps about the bulb fields. One enjoys their gaiety and the feeling they convey of life bursting through at the end of winter, the other thinks the whole thing a little unsophisticated, all right for the Women's Institute annual trip but not suitable for the discerning traveller. On this issue, I find myself at one with the WI. I like the bulb fields. My attitude is no doubt influenced by my having once spent several months working for the grand old bulb firm of van Tubergen in Haarlem. The smell of hyacinths in bowls at this time of year always makes me long to go back.

That said, I do understand that the bulb fields are not everyone's cup of Douwe Egberts. The same may be said of the public bulb gardens, though they are a must for any WI itinerary. Most famous is the Keukenhof at Lisse, a garden made in the grounds of an old estate (Keukenhof means 'kitchen garden') owned by the Countess of Holland in the fifteenth century. Here bulb firms rent space to grow their flowers in beds in grass, among trees, shrubs and statues. My reservation about the Keukenhof is not so much that commercial considerations make the layout fragmented and incoherent (the bulbs appear too much like a rug woven by a committee at loggerheads), but rather that the bulbs, which never grow taller than three feet, are dwarfed by the trees. The lack of scale seems more obvious in an undulating garden of curving paths and water 'features' than it does in the flat fields under the arc of the sky.

The Keukenhof is nevertheless an unforgettable spectacle, and an excellent showcase for the industry. You will not see anywhere else such a range of cultivars grown, many of them, rather frustratingly, difficult to find in Britain. For example, last year there was an exhibition in the Queen Beatrix Pavilion of a dozen varieties of the so-called 'split corona' or 'orchid' daffodils, the result of breeding work done mainly by Gerritsen of Voorschoten. Some, but by no means all, are available from large bulb firms in Britain, but they are practically non-existent in gardens as yet. These daffodils have been bred with trumpets (coronas) which are splayed flat against the outer petals, a diversion from the norm which is, it must be said, sometimes more striking than pretty. Perhaps they do not look sufficiently like our idea of a 'daff' ever to be popular here.

Those who consider themselves discerning should also visit the humbler Hortus Bulborum at Heiloo, outside Limmen, where no attempt is made at landscaping at all. This garden simply consists of a field containing rows and rows of museum-piece bulbs, most of which are tulips. None is any longer in commercial production, which makes this garden an important source of genetic material for breeders. Early tulips were short-stemmed, fat-budded, by no means always striped, and just as striking as modern varieties. The oldest bulbs in this garden are the red and yellow versions of 'Duc van Tol', which date from 1595 and figured in the Tulipmania of 1637. Here in the grey earth is written the history of the Dutch people's four-hundred-year love affair with tulips.

To visit Holland in the spring is to become aware that the tiniest suburban garden there boasts a greater variety of flowering bulbs than is commonly seen even in large gardens in Britain, where a few snowdrops, 'daffs' and bedding tulips are all that most people aspire to. Many in Holland turn their gardens over almost completely to spring-flowering bulbs. This can, of course, pose a problem when the summer comes and the flowers and foliage die down. There is no difficulty for the Keukenhof, which closes its gates to visitors in late May, but private gardeners must resort to planting large quantities of half-hardy annuals – begonias, busy lizzies and petunias – to fill the gaps. The WI probably like that, but I am not so sure.

5

Practical Gardening

Jobs not to Do   *6 April 1985*

The enjoyment of life, as someone must once have said, depends upon coming to terms with the gap between aspiration and achievement. Most gardeners' love of flowers far exceeds their enthusiasm for gardening. Their zeal is undermined by a sluggish metabolism, stiff joints, shortage of spare time, and other symptoms of a lack of inclination. The guilt that results is made worse by the welter of information provided by many gardening magazines and newspapers under the heading of 'Jobs for the Week'. Such is the formidable catalogue of activities advocated that the average gardener feels defeated at the outset. The comprehensive nature of these columns is undeniable, the industry that has produced them commendable, but their usefulness (except in finishing off the faint-hearted,) is doubtful, because they rarely mention how important, desirable or difficult is a particular job. Disbudding greenhouse perpetual carnations is referred to in the same even and unemphatic tone as making seedbeds for hardy annuals, spraying against cherry slugworm recommended in the same breath as watering in dry weather.

While not wishing to underestimate the benefits of pursuing excellence, especially in something so unimportant, I have sympathy for the diligent beginner who is taught to believe that gardening consists largely of seeking out and destroying beasts she has never seen and would not recognise if she did. Where is the fun of that? If all she feels she can look forward to is a lifetime of fumigating, scarifying, spiking, weedkilling and cleaning up, why should she ever begin?

As with housework, the gardener should sometimes ask herself what it is all for. If she takes pride in the continuous succession of jobs that she does, all well and good. If, however, obeying every injunction destroys the peaceful enjoyment of her garden, then she should appreciate the charm of roses growing at the top of leggy unpruned shrubs and be satisfied with sufficient quantities of fruit from unsprayed and unfed apple trees.

If I were charitable (which I am not, for one is always hardest on one's own kind), I would say that horticultural journalists are merely anxious to cover every aspect of a huge subject in a limited space. Anxious, that is, to inform and enlighten a public as numerous and diverse as pebbles on a beach. But their imaginations fail to rise to the challenge of encouraging rather than overwhelming. Furthermore, in the process of reduction, the meaning often becomes obscure. Who but an experienced viticulturist could make anything of 'tie rods of spur-pruned vines to training framework and replacement rods of Guyot-pruned varieties horizontally . . .'?

If I am vehement it is because I have been at fault, and now have all the self-righteousness of the Born-Again. It was I who for some years burdened the spare hours of East Anglian news-paper readers, every other Thursday, with a seemingly endless list of jobs to do. 'Tell it not in Gedney Drove, publish it not in the streets of Walsoken' should have been my watchword – or not, in any event, with such mind-numbing completeness. Nothing was omitted, for fear I might be supposed to have for-gotten or, perish the thought, never to have known. The problem was particularly acute in winter: fewer genuine activities, but just as many column inches to fill. 'Fix greasebands round fruit trees' the bald advice would run. What, how and, most particu-larly, why? I did not elaborate. Never a word about how inevit-ably greasebands stick to your fingers, as fast as newly chewed gum unhappily discovered under a desk-lid. I once, and only once, spent an autumn morning fixing these bands to the trunks of apple trees; we became as inseparable as Laurel and Hardy.

What no one will say, and amateurs must find out for them-selves, often painfully and slowly, is that as much as anything else the beauty and pleasure of gardening lies in pottering (an occupation open to the owner of the smallest garden), espe-cially in that blessed hour after dinner in the summer when nothing more energetic need be contemplated than training clematis tendrils round supports, dead-heading roses, or sniff-ing the mock-orange-scented air. In all this lies the true enjoy-ment of gardening, not in the headlong succession of often hard, futile or needless chores, with all the guilt of failure, incomprehension and opportunities missed to bear one down.

So at the start of a gardening year that will be more than filled with necessary work, here are some jobs on no account to do. Do not dig up your rhubarb crowns in winter in order to force them into early growth in the greenhouse, for rhubarb has a season that is quite long enough already and only an oxalic acid freak would wish it extended. Do not tip-layer cul-tivated blackberries in summer; they will do it quite well of their own accord, always providing that you have failed to tie the shoots to wires the autumn before. And do not pollinate your greenhouse tomatoes with a rabbit's tail or paintbrush; being self-fertile, they are perfectly capable of managing the business themselves.

Stamping out Slugs 12 January 1985

I have sometimes wondered what strong inner compulsion drives gardeners to struggle outside in early winter, equipped with rakes and besoms, to clear the shrub and flower borders of leaves and weeds before the coming of the hard frosts. Never mind that this annual task, often accomplished in the face of a biting, rain-heavy wind, means the removal of weeds already seeded and dying, and leaves that would rot where they lay and be pulled by hardworking worms down into the soil to make humus. Is it really to prevent the leaves from flying about? Or is it just obsessive neatness, like over-zealous house-work – pointless, but deeply satisfying? Perhaps putting the garden to bed in the winter is an emotional response to the hurrying of time which also acts as a talisman against disaster, like tidying one's underwear drawer before going on a long journey. It may also engender the feeling that Fate will smile on us next spring if, like the Wise Virgins, we are up and about now, topping up our vessels like anything.

Then, as I weed in the wet leaf litter (for I am not immune to this yearly urge), I feel a sudden and convulsive surge of dis-taste as my fingers curl round a soft and clammy garden slug, and I remember how we justify all the raking and clearing: to leave the slug no hiding place.

It is hard for the non-gardener to appreciate fully the depth of animosity felt by gardeners towards the slug, or with what heart-hardenings it is possible for the mildest and gentlest of women to kill them without compunction. I myself favour the stamp and the sharp twist of the Wellington boot heel as being the kindest and quickest method of despatch. I have become adept and, though prone to self-reproach and eager on Sundays to add the wilful and malicious murder of slugs to the confession of all my other manifold sins and wickedness, I know in my heart that I am not truly repentant. The mental pictures that spring unbidden to my mind of burrowed Persian bulbs collected by friends in the Elburz mountains before the Revolution, sawn-off shoots of named delphinium varieties and punctured hosta leaves is enough to confirm me in my transgressions.

Surprisingly, there are as many as a dozen species of slug common in gardens. The real rotter is the netted slug, because it is so fond of green vegetables; so much so that between it and the terrific hordes of white fly encouraged by this year's weather I have resolved to give up growing cabbages altogether. This is followed in criminality by the common garden slug, particularly found burrowed into maincrop potatoes in damp autumns.

It would be foolish, however, to condemn all slugs out of hand. The great grey slug is a noble mollusc that is not especially unsympathetic to the designs of the gardener. One occasionally saunters – there is no other word for its arrogant and, even in slug terms, unhurried progress – across the paving outside the kitchen door. I stay my hand, or rather boot, if only because in order to mate this creature goes to the bother of climbing a tree and hanging, wrapped in an embrace with its paramour, suspended from a thread of mucus. It deserves to be left in peace.

For those readers too squeamish to contemplate the method of slug liquidation that I employ there are as many other ways of killing them as there are old boys in the pub prepared to exchange advice for a pint. The chemical approach, favoured by those who do not have pets or garden birds to be poisoned, involves the use of slug pellets. These work by dehydrating the slugs, which must result in a truly agonising and, in wet weather, slow death. It does take the guilt to one remove, however, for those whose stomachs are too weak for the direct approach. If used, pellets are best used really sparingly, placed under a clay pot that will stay put, where other animals cannot find them.

Popular these days is the pot or yoghurt carton full of pale ale. This technique takes advantage of the gastropodal passion for beer. The idea is that, attracted by the smell, they will topple headlong to a happy drowning death. If the number of slugs usually to be found floating keel up the next morning is anything to go by it is very efficient and ecologically most considerate, or it would be if plastic cups rotted.

All this talk of killing makes me feel a little ashamed. Perhaps I should mend my ways and return to the old remedy of describing a circle with wood ash or clinker around particularly favoured plants. The sensation to the slug must be similar to that which we would experience if we crawled naked over peagrit gravel: unpleasant and certainly a deterrent, but not fatal.

A winter clear-up to discourage slugs may be wasted labour anyway, for some naturalists believe that if the leaves are left to die on flower beds the slugs will eat them rather than the new green shoots of cultivated plants in spring. I wonder. What I do know is that, slugs or no slugs, the mainspring of our action is restlessness. Cleaning the borders gives us something to do. Gardeners get fidgety in the winter, probably because there is always far too much to do in the other seasons. Caught up in perpetual motion, we cannot stop. Each Sunday, bolted apple crumble forms a hard lump somewhere near the sternum as we dash out to catch every precious minute of daylight. How can we be told there is little point? It would be as cruel as stamping on slugs.

Bucket and Spade   *10 August 1985*

The stamens of the philadelphus have finally browned, the last pink petal of 'Madame Grégoire Staechlin', that great fat blowsy tart of a rose, has fluttered to the ground and the height of summer, such as it was, is now past. I do not especially regret it, for gardeners waste too much time looking back on past glories, and I learned long ago that it is quite possible to have something flowering in the garden in every month of the year, if only rainswept and tattered early crocuses.

Thoughts of a holiday, however, cannot be by many days delayed. Unfortunately. For me, as for many gardeners, holidays are anathema. They are invented to confound us by people with too much time on their hands. Holidays take me from the ripening tomatoes and maturing runner beans and the blossoming of some of the best clematis. Will I ever see my *Hoheria lyallii* flower? I must work so hard before I go, not only to discourage the sow-thistles from setting seed but simply for my pride's sake. I hope the birds, having eaten the raspberries, will enjoy the sight of my orderly garden, gradually blurring into muddle again before I return. I worry about the watering, inside and out, and am forced to take part in complicated bartering of mutual dependence with my neighbours. One year, in return for a tended vegetable garden, I promised to feed some trout in a small fish-farm. Twice a day, my heart choking me and weighed down almost to the point of immobility by buckets of fish pellets, I launched myself across a bull-infested field.

I, who am quite content to stay at home, have even been cajoled into going abroad to Spain, where I lie on a barren beach, restless and unhappy with nothing to look at but dusty palms and the ubiquitous oleanders and bougainvillaea. Sour envy is all I feel at the size of the yuccas near the Bar Manchester Arms at Cala Millor. If I am lucky I can escape for an hour from the business of browning, or rather pinking, to roam the hills behind the beach collecting cistus seed.

One Easter, in Sussex, in an excess of gratitude towards those who had lent us a house, we decided to plant their empty garden on the seaward side. I wish we had inquired more closely as to why, except for an indifferent stretch of moss, it was empty. I know now that such a generous impulse should have been strenuously resisted. I did indeed entertain misgivings, but suppressed them lest they be misconstrued as idleness and ingratitude. This garden had only a low wall between it and France and, we should have realised, was battered by unremitting salt-laden winds. The unforgiving soil, shallow, alkaline and knobbly with pebbles, was, I suspected, dangerously salty from the frequent floodings it received in winter. Nor had we properly bargained for the owners' fierce attachment to being able to see the sea, or at least that part of it that meets the land, which made the planting of any sort of shelter belt unacceptable.

We planted everything we thought would like to grow: pinks, hebes, hydrangeas, tamarisk, fuchsias, potentillas and olearias. We watered and watered desperately and, as it turned out, vainly, to try to wash the poisonous salts out of the soil. By the summer holidays, a few months later, there was little left except some redoubtable pinks and hebes, *Rosa rugosa* varieties which I truly believe to be indestructible, and one or two of that most boring of shrubs, *Griselinia littoralis*, with leaves as shiny and inanimate as a billiard ball. The hydrangea leaves were so scorched that they looked as if a match had been set to them.

We had fallen painfully between the two ideals of shelter and view. They are mutually exclusive except where the seaside-dweller, preferably equipped with yachting cap and mounted telescope, takes to living upstairs. Without a good windbreak there are few plants, even of the tough and leathery kinds, which have more than an even chance of becoming established.

The news from the coast is by no means all bad, however. A little way from the sea, where there is the shelter of walls, seaside gardens must be the greatest fun to cultivate. Tenderish shrubs like escallonias and pittosporums, that drag out a bare existence in frosty inland gardens, grow like nettles on a dung heap, petunias rival cabbages in vigour, and you can take no very daring chances with palms and agaves, lapagerias and yuccas.

What is more, the seaside gardener has on her doorstep, in seaweed, one of the best natural sources of potash known to man. It is especially useful for fruit, and potatoes love it. If you can bear the smell, and picking out the man-made rubbish, and can fight off the flies which buzz around it incessantly in summer, you have a manure that, when dry, will rot down rapidly in flower bed or vegetable garden. For those who can contemplate cheerfully the fetching and carrying, the rewards are rich. I know I shall be packing my bucket and spade.

The Answer Lies in the Soil   *11 October 1986*

It is the common lot of all gardeners for each one to think him- or herself uniquely unfortunate. Misfortune takes many forms, which conspire to confound the best-laid schemes and make the creation of an attractive garden little short of miraculous: there is the north-facing aspect of the garden (which, except when actually perched on a windswept hill, is invariably a frost-pocket); the deep shade cast by huge, thirsty trees, usually belonging to unhelpful neighbours and the subject of Tree Preservation Orders; the everlasting weeds; and, first among equals, the awfulness of the soil.

How often have you met a keen, hardworking gardener who boasted of their fertile, friable, slightly acid, moisture-retentive medium loam, workable in winter yet never droughty in summer? Quite.

I must admit that I garden on quite a good soil, a sandy loam,* but let me hastily assure you that it is not perfect, for it dries out too quickly in summer. Nevertheless, it is a village soil, roamed

*This was written before our move (more fool us) to a garden on heavy clay.

for many years, I imagine, by the family pig and therefore quite fertile for one so quick-draining. There is, I know, something a little vulgar, certainly in questionable taste, about owning up even to that much; it smacks of a desire to grind the faces of other gardeners into their stiff, unworkable clays.

I am one of the lucky ones; but however legitimate the complaints made by gardeners about their soil, it is important that they do not turn away in unhappy resignation and concentrate their efforts on lovely flowers alone, for that loveliness has a great deal to do with the dirt in which the flowers grow. I rather admire people who embrace the subject of soil with enthusiasm, who feel at home with iron chelates and the colloidal nature of humus and will say without hesitation, when asked where they garden, 'We live on a band of greensand, overlying gravel, at a pH of 5.5.' I consider it a useful accomplishment to be *au fait* with the constituents of one's soil; not to be be boasted about, of course, but a source of quiet satisfaction, like knowing where to find the Travellers' Club without asking a taxi-driver.

A simple kit or meter will serve to test the acidity or alkalinity (pH) of the soil with acceptable accuracy; this is a facet of the subject about which no gardener should remain in ignorance, for some plants like rhododendrons are most particular. There is not, in the long term, a great deal we can do to alter the pH so that we may grow plants which do not like what we have, nor should we bother overmuch, except by adding lime to an acid soil to improve the disease-resistance of cabbages. After all, the pH of the soil, and the plants that it supports, can be as firm indicators of place as vernacular architecture, and should not be transformed with the chemical equivalents of uPVC replacement windows and 'stone' cladding. Fortunately, plants know their place as well as any Victorian parlourmaid, so few calcifuge plants will survive healthily for long in a limy soil, however much peat is dug in to try to acidify it.

That said, it is possible to do much to improve the structure and, to a lesser extent, the texture of a soil. Texture refers to the proportions of the different-sized particles of sand, silt, clay and organic matter in the soil. A preponderance of large particles makes a light soil, and small particles a heavy one. Structure refers to the way in which the particles are aggregated together into 'crumbs'. A soil with good structure has large crumbs, so that there are wide pores for oxygen and water to pass through. A sandy soil, consisting of many large particles which allow water to drain too quickly, is made more retentive by frequent digging-in of compost or manure. Thin soils over chalk require the same treatment, and the subsoil should be forked over to enable the roots of plants to reach down to any available water. Even a clay which seems as difficult to get through as an episode of *Brookside* can be lightened in time by the persistent digging-in, in autumn or winter, of copious quantities of grit or sand, ashes and well-rotted compost. If that is too hard, simply lay organic matter on top and leave the rest to the worms. There is comfort even in despair, for clay soils hold nutrients and water well, and are therefore very fertile.

One may dedicate oneself to lesser things in life than the improvement of one's soil. It is honourable work, intellectually absorbing and physically tiring. A really heavy clay is certainly not to be wished on one's worst enemies, or not the older ones among them, but there are plenty of fine gardens made on the stickiest substrate. The famous garden belonging to Beth Chatto at Elmstead Market in Essex, for example, has been developed over the last thirty years on a soil that was in places too heavy and waterlogged to be farmed. There is a sloping bank of pure clay, formed when a lake was excavated, which has been transformed into a successful flower border by the piling on top of masses of compost. Counting up misfortunes is agreeable, but anyone who has seen this garden may prefer in future to save their breath for the digging.

Agony Aunt   *29 August 1987*

This month, having run out of ideas of my own, I answer some readers' letters.

> We have some hard landscape in our gardens which consists of a series of stone steps down which water flows in a cascade. We are worried that algae growth may build up.
>
> A.D., Chatsworth, Derbyshire

I'm afraid you don't tell me how big the waterfall is or how much time you can spend in the garden (or, indeed, whether you are lucky enough to have any help!). Can I suggest that once a year you turn the water off and give the steps a really good scrub with a stiff brush and some diluted household bleach? You'll need to use some elbow-grease! I wonder if you've considered covering these steps with a length of blue butyl pond-liner? You can get it these days at most garden centres. That way your waterfall will be easier to maintain and could become a real feature in your garden!

> I've got a plant which was given to me for my birthday; it is in a pot; it may have had flowers but I can't remember seeing any and the leaves have all turned brown and dropped off. Have I over-watered it?
>
> A.W., Spalding, Lincolnshire

The plant you describe is quite obviously *Dieffenbachia picta*, which was introduced from Brazil in 1820. It is a tropical plant which requires a minimum winter temperature of 16 °C (61 °F). It has dark green ovate leaves, blotched with white, and these easily go brown and die off in dry conditions. So retain high humidity at all times, without over-watering, by standing the pot on a tray of pebbles which can be topped up with water. West Indian plantation owners used to give the leaves of this plant to recalcitrant slaves whose mouths, when they ate it, would swell up painfully. Don't, however, give it to the *au pair*, or she'll never finish that language course!

> Can you recommend a really good spray to deal with greenfly?
>
> R. B., Bracknell

Alas, not yet. But I live in hope that a chemical company may one day offer a sufficiently large retainer for me to do so with complete confidence.

> I was having an argument with my neighbour over the fence last night. He is adamant that Canadian pondweed, *Elodea canadensis*, is an example of agamospermic repro-duction. I maintain it is just good, old-fashioned apomixis. Surely it couldn't be kleistogamy, could it? What do you think?
>
> D. B., Durham

I don't.

> When we moved here, we found that part of the garden contained nothing but plants which are white. What can you suggest to turn this drab corner into a really lovely splash of colour?
>
> N.T., Sissinghurst, Kent

Well, don't some people have extraordinary ideas! And the oddest gardening opinions always seem to be held by anyone from whom one buys a house! Fortunately, your problem is easily put right – just think yourself lucky that you haven't been saddled with some unsightly permanent feature (say, a huge brick tower) in the middle of your garden! May I recom-mend what I like to call my 'Bobbydazzler Selection', which is guaranteed to brighten up the dullest spot? Throw out all those dreary white tobacco plants and Madonna lilies, and replace them with a smashing mixture of busy Lizzies (like *Impatiens* 'Novette' in deep orange, scarlet-red, salmon, white and rose), some of the new, fringed, double petunias and lots 'n' lots of colourful 'Giant Cactus-Flowered' zinnias. That's sure to create a sensation should you have any visitors. But don't go too far. Restraint can be a virtue in gardening. Why not keep the hostas?

> I have a back garden which is shaded by a large sycamore tree in the next-door garden. The soil is a mixture of light stony brash, yellow subsoil and dried-up cement which the builders left. All the local cats use it as a public convenience and water drips from an upstairs overflow onto it, turning it into concrete in summer and a mud morass in winter. I should like to grow colourful hardy flowers in it for twelve months of the year, in a simple, tasteful harmony of pinks, blues and greys. It should not cost much to plant up, should last without alteration for years and years, and must be labour-saving. What can you suggest?
>
> E. B., Clapham

Nothing.

> What is the Fibonacci series?
>
> U. B., Northants

I could not even begin to guess. Anyway, it is very bad form, even if you haven't received enough letters, to write to yourself. Look it up.

Miss Buchan regrets that she is far too idle to enter into private correspondence.

'Give Us the Tools . . .' 10 December 1988

Someone who shall be nameless, but who knows quite well who he is, has broken my gardening knife. He borrowed it for a trifling task and succeeded somehow in breaking the hinge. The blade now lolls like the head of a maltreated doll, and I am not very happy.

This knife has a wooden shaft and a curved blade; I bought it fifteen years ago in the ironmonger's shop in Peebles. The first day it was used, before I was accustomed to its ways, I shut a finger in it, not realising how fierce was the (now sadly shattered) spring. The blood flowed in an unnerving stream, to the concern of the gardeners with whom I worked. Since then, it has been my constant companion in the garden, used for cutting string, opening bags of compost, chipping sweet pea seeds, digging weeds out of the lawn, and stabbing the occasional slug. And now it's bust.

The reason I feel so pained is that well-worn tools like that knife can inspire almost as much affection and respect in me as our long-established and fruitful walnut tree. They become part of my gardening landscape and, if anything, are the dearer for being so homely. The older they are the more they seem moulded to my particular needs. I feel the same about an assortment of elderly, inherited, forged-steel border forks and spades; although heavier than their modern counterparts they seem, with their triangular and (it has to be said) worm-eaten wooden handles, to come to hand more easily and smoothly than the modern plastic-moulded ones do. The spade blades have been sharpened over the years so that they cut through the soil like wire through cheese. So fond of them am I that I feel no desire to own one of their epoxy-coated successors.

In any event, I have no room in the tool-shed for anything more, so full is it of one-use tools which never see the light of day except on the one winter's afternoon in the year when I tidy out the shed. Just as I rarely use that special onion slicer that seemed indispensable and infinitely desirable in the hands of the county show barker, so that clever little gadget which sows seeds evenly gets about as many outings as a spavined racehorse.

Fortunately, such an embarrassment of riches means that I need not bother my little head about anything so potentially problematic as the 'multi-change tool system'. Several manufacturers now make handles on to which can be clipped a variety of different tool heads (although not all at the same time, you understand). Wolf Tools are particularly noted for introducing heads to satisfy needs you never knew you had. These 'combisystems' are, however, ideal for beginners who have as yet no tools at all, provided they do not feel honour-bound to buy every single head.

When Christmas approaches and with it at least the possibility that someone might think to ask you what you want, it would be as well to have handy a short list of essential hand tools. My list comprises: knife with a straight blade and a wooden rather than plastic handle (the latter splits, in my experience); oilstone on which to sharpen the knife; a pair of secateurs, as expensive as your family's affection for you will allow, with a scissor rather than anvil action; pair of handshears; curved pruning saw; long-handled loppers (optional if you have no well-grown shrubs and trees); digging spade; digging fork; border fork; shovel; hand-trowel and hand-fork; garden 'line'. For leaf-raking or for clearing up the lawn edge after you have been weeding, I suggest the rubber rake – or besom, if you prefer; either is essential for doing the all-important 'proper job'. Only if you have vegetables or grow your bedding plants in straight lines will you have much use for a Dutch hoe, and then it should have a small head. The same goes for the draw hoe, which is unrivalled for earthing-up potatoes and taking out a seed drill but completely useless for anything else. Some people swear by the three-pronged cultivator for border weeding, but I much prefer a border- or hand-fork. Occasionally you will need a half-moon spade, to cut a straight edge to the lawn, but a sharpened ordinary spade will do almost as well. A trug basket is useful for carrying around your hand tools, and a pair of kneeling pads essential.

I cannot deny that I do sometimes dream of the day when I shall own a set of stainless steel tools. They are unalloyed pleasure. The blade of the spade enters the ground cleanly and with little or no resistance, yet is effectively self-cleaning – unlike my forged steel spade, which must be scraped clean of mud after use and rubbed with an oily rag to protect it from rust. The reason why I do not own such a set is simple: these tools cost about three times as much as ordinary garden forks and spades. However, this year it might just be worth my while asking Someone for them for Christmas; after all, generosity and contrition do often go together.

Beastly Beatitude   *18 August 1990*

It is quite common for people to turn to me at parties and say: 'Of course, I'm never happier than when I'm gardening, and I think gardeners are *such* nice people, don't you agree?' Because I am a gardener and therefore, by this analysis, a nice person, I do not reply, as I should like to, that niceness does not make them good gardeners; in fact, it is a positive disadvantage.

The underlying assumptions, I suppose, are that gardening, being one of the arts, both necessarily has a beneficial effect on those who do it and attracts a certain rather wonderful sort of person. Rudyard Kipling did nothing to nail the first lie when he wrote 'The Glory of the Garden glorifieth every one'. As for the second, I can think of at least one appalling stinker who was absolutely devoted to the works of Wagner.

In fact, in my experience, the beastliest people find gardening comes very easily to them. It is just the job for bullies, sadists and supremacists. Stamping on slugs, trapping mice, squeezing aphids, unnerving birds with cotton strung above the polyanthus, throwing unwanted but perfectly healthy plants on the compost heap, grubbing up plants which are in the wrong place – these are tasks from which the good gardener cannot shrink.

I should not wish to go so far as to say that gardening has an actively brutalising effect on those who practise it, simply that the more mean-spirited or, certainly, hard-hearted you are, the better you can do the job. Those of us who are uterly wets and cissies, the fotherington-tomases of the horticultural world, never do as well as we might because we are simply not ruthless enough.

Take one example of many from my own depressing history. I have an unusual hoheria planted against a sheltered west wall. I put it in eight years ago to provide some 'colour' in August. I went outside today to look at it, in the confident expectation that, as usual, it would be a dull green lump, only to discover that it has begun to flower. Two exceedingly and exceptionally mild winters and hot summers have finally shamed it into action. For an instant I congratulated myself on my restraint, which was in fact simply an ignoble mix of cowardice, idleness and parsimony (the 'I've spent good money' syndrome). Then, having taken a more considered look, I realised that the flowers were really not so spectacular as I had been led to believe: small, off-white, scentless, sparsely borne. (I bought it on the recommendation of a gardening writer, something one should always be wary of doing.) Now I have not even the excuse that I am waiting for it to flower. But I bet it will still be there next year.

What brings out the truly soppy in me is the unwanted rooted cutting. We are taught, rightly, to take more cuttings than we will need, in case there are failures. So I am nearly always left with too many, and I grow them on because I cannot bring myself to throw them away. I cannot bring myself to look after them properly either, so unless some kind friend comes round the garden they have to wait for the garden fête, by which time they are too starved to be acceptable.

All this is a common failing of garden owners. I wish I had a euro for every time someone said to me 'It ought to go but I have not the heart to throw it out.' The trouble is that they have too much heart and not enough head, and the results are shambolic borders and overgrown, over-mature plants. What is more, they expect the rest of us to cheer such sensitivity.

So don't listen to all this tosh about niceness; it will get you nowhere. If you are not blessed with it already, learn to cultivate a little heartlessness. Your friends may give you the cold shoulder, but you will have a much better garden. And I will be glad to talk to you at parties.

Don't Do It! *17 November 1990*

There are two informative booklets currently available which fill me with gloom. They are called *Fence It* and *Pave It*, and they are the successors to *Strip It*, *Paint It*, *Paper It*, *Tile It*, *Shelve It*, *Frame It* and *Wire It*; not to mention, if you can say it without swallowing your tongue, *Mirror It*. There is one still to come, which shows you how to hammer nails into walls – called *Sod It*. These books are all part of a series called *Do It!*, a title which seems to me ill-advised – not because of the no doubt unintentional *double entendre*, but because of the bossiness inherent in the '*!*'.

I can just imagine how such a title came into being. During a brainstorming session at the publisher's Geoff, the ideas man, came up with a really great idea. 'Hey,' he said, 'I've got this really great idea. Let's kick it up the flagpole and see if it gels.' But all Geoff has succeeded in doing is putting me off.

Not that that is very difficult. I have never Done It much, and I suspect that the 70 per cent of gardeners in Britain who happen to be women do not Do very much of It either. That may seem an intolerably sexist remark, but I have yet to meet a woman gardener who really revelled in DIY. It is the fortunate exception, among my generation at least, who managed to wangle her way into woodwork at school. Most of us have a morbid fear of the drill, the dolly and the awl.

Which is why I watched in horrified fascination the other day as I was shown, by the presenter of a gardening programme, just exactly how I might construct my own pergola. It was easy, apparently, as long as I was prepared to take a little trouble. My negative reaction was the stronger because I have spent a good part of this summer watching a small team of able builders laying paths, building retaining walls and erecting trellis in my own not very large garden. The amount of hardcore whacking, cement mixing, stone and earth moving and pavior cutting which has gone on suggests to me that this sort of thing is hard, physical and time-consuming labour, requiring a considerable degree of skill. How could I, with only the odd afternoon to spare when I am not gardening, be expected to produce anything similar?

To be fair to the *Do It!* booklets, there are lots of jolly explanatory diagrams featuring chaps in gumboots hammering posts in the ground. But this just cruelly reinforces the reader's impression that this kind of activity is both simple and desirable. It would be far kinder, in the end, to tell us how extraordinarily fraught and even self-defeating the whole business can be.

An estate agent friend of mine says he always knows when he is showing a potential customer around the house of a DIY enthusiast because the wardrobe doors come off their hinges as he opens them. If not done well, home 'improvements' can look awful and are certainly unserviceable. The same is true in the garden. I have known several people who have laid paths with old bricks in their gardens which have looked very pleasing initially. However, because these gardeners have not known whether the bricks were frost-resistant or not, the first hard winter has seen these paths flake like almonds.

Left to myself this summer, I would have been lucky to have produced a simple stretcher-bond path, laid on sand, which did not wobble extravagantly underfoot. As it is, I now have something approaching the Lower Terrace at Bodnant in miniature. *Folie de grandeur*, perhaps, but even the plainest and smallest piece of landscape work takes more skill and time than most gardeners care to lavish on it. That being the case, commentators should come clean and admit that, if you want pleasing, finished results, you will probably have to pay for them. There are times when good money has to be spent, and no amount of upbeat explanation about how to gouge out your mortises will alter that.

Chocs Away! 19 October 1991

I am not a chocoholic. I can go all the way till lunchtime without so much as a square of Dairy Milk. If I want to. I just do not see anything wrong in eating something which only gives you spots and makes you fat. I feel guilty only because of the blinkered attitude of other people. A liking for chocolate is so widely considered a weakness that the only time I have not felt shamefaced about it was during pregnancy, when I could make out a convincing case for the iron it contains being good for me. Is it any wonder, therefore, that I seek consolation in my garden, where I can indulge my chocophilia without fear of censure?

It is not possible, of course, even in these strange climatic times, to cultivate one's own cocoa trees outdoors in the East Midlands, but one can grow the next best thing, a rather fetching tenderish perennial called *Cosmos atrosanguineus* which smells deliciously of chocolate. And not just any chocolate, either, but that luxurious dark chocolate from Terry's, which comes in a red wrapper.

In summer and autumn *Cosmos atrosanguineus* produces two-foot-tall stems of single flowers which are the colour of a damson with its bloom rubbed off. Even the leaves are attractive – for a member of the daisy family, at least. The only blot about this particular plant is that it is not really winter-hardy. However, the tubers may be dug up each autumn and stored until the following late spring, as if it were a dahlia.

Until this year, this cosmos was my main chocolatey solace in the garden. However, patient searching in Thompson and Morgan's seed catalogue last winter revealed another possibility, *Berlandiera lyrata*. The seed of this half-hardy perennial, which is incidentally in the same family as the cosmos, was sown only in March, yet the resulting plants managed to produce a very long succession of drought-resistant, chocolate-scented flowers this summer. These flowers consist of yellow ray petals and maroon centres, and are held twelve inches above slightly coarse, silver-green leaves. It has to be said that the chocolate fragrance is not my favourite kind – it falls somewhere between Galaxy and a stale Christmas-tree decoration – but what it lacks in quality it makes up for in strength, particularly after rain.

Yet welcome as this new addition is, it has turned out to be no more than a taster. For, this season, a product has burst upon the horticultural scene which is far more exciting. This is cocoa shell, the waste product which remains after the 'pips' of chocolate have been removed in chocolate manufacture. It has become, for me, the Charbonnel et Walker of mulches. Cocoa shell has, apparently, been used for many years as a humus-providing, nutrient-rich soil conditioner and mulch by those living near chocolate factories, without the rest of the world having any inkling of its benefits.

Derek Spice, whose idea it was to market the stuff, under the name 'Sunshine of Africa', is a chocolate man through and through. He once worked in the industry, his brother makes chocolates, and his father was a cocoa importer who spread nothing else but cocoa shell on his garden for fifty years. Cocoa shell is not particularly cheap, for there is competition for the supply from cattlefeed manufacturers, but I am told that less is required each year to have the same beneficial effect. I hope for Mr Spice's sake this is not literally true.

Cocoa shell is higher in the major nutrients than farmyard manure and is a great deal easier and lighter to handle, although it can become rather slippery when wet. What is more, the shells do seem to act as a deterrent to slugs, both initially when they are sharp and also after they rot down to form a porous but weed-suppressing blanket. This is because the mucilage and gums in the shells, which are released by rainwater, make the mulch surface too tacky for them. There is some anecdotal evidence that cats do not like cocoa shell either.

The only real disadvantage of this material, and for me it is a serious one, is that once it has degraded to an attractive chocolate-coloured humus, it loses its delectable scent entirely. Oh dear, choc's away.

Call a Spade a Brotzman   *7 October 1995*

Until recently, I have always gardened on a light, free-draining, sandy soil. I used to complain about it ceaselessly because it always dried out by August, whatever the summer, so that flowering faltered and leaves withered. When we moved house and bought a garden on a heavy, fertile, moisture retentive stony clay, I was paradoxically blithe, keen to reassure doubters with homely sayings like 'A clay soil breaks your back but not your heart'. Now, however, I cannot easily draw the anatomical distinction.

That is, since the day I planted pheasant-eye narcissi in grass in a proto-orchard. The idea was to grow them in broad strips, to make mowing easier, and I looked forward to spring when they would wave and dance under the fruit trees. The future charm and timeless beauty of the scene enchanted me. I had ordered, from a wholesaler, a satisfactorily cheap 25-kilo bag of bulbs. That is 56 pounds in weight in old money, or, to put it more simply, about 1,500 bulbs. I was not counting. What I wanted was generous, full-bodied planting.

It was early September, before the rain had softened more than the top inch of soil. I was impatient, though, because such large-scale planting could well take several hard-won after-noons. In any event, I scarcely wanted to wait for the transition from concrete to glue which happens so quickly to a clay soil in a wet autumn. Clay is not the ideal medium for daffodil bulbs, I know, but I had plenty of grit ready to put into my trench to help drainage. I embarked, cheerfully, on the task of digging a long, deep, broad trench with the Brotzman Memorial Spade.

This spade (called a 'shovel' in American) is named after a friend, a conifer nurseryman who, with the good-natured thoroughness which so characterises Americans and shames more recessive Brits, had once brought it as a present, wrapped up carefully, in his luggage. The blade, sharpened by long and abrasive contact with Ohioan sand in the hands of a massive muscular Ohioan, is as square as his beard. The haft is wooden but almost entirely encased in forged steel, and bears the legend 'A.M. Leonard and Son, Piqua, Ohio. Heat Treated'. Although almost twice as heavy as any other spade I use, it is well balanced and comes to hand easily. If anything could make the work light, it was the Brotzman spade.

But stony clay is not like any other soil. When it is without moisture, which is the case after a long drought, you can only dig a trench by making a slit and then progressively shaving the side of it, removing slivers of compacted dirt as thin as milk-chocolate gratings. You dare not chance digging with vigour, for the shocks which go up your arm when spade meets subterranean stone wear down the spirit and, anyway, those stones effectively bar the spade's progress.

It was no good. I would have to use the pickaxe, a tool which I have always considered suited only for husbands wishing to remove tree-stumps. Nowhere, ever, had I seen or heard anyone suggest that you might have to plant bulbs with a pick.

In the end, I used a mixture of the spade and pick, preferring to stand the jolts and stops and the risk of gashing my leg rather than those unnerving, unpredictable shocks up the arm, but turning to the spade once the biggest stones were removed. The delicate pure-white-petalled, orange-eyed, scented flowers no longer danced before my eyes.

Although many gardening tasks can be done badly and yet still be effective, I knew that if I skimped this one I would destroy the reason for doing it. Bulbs planted too shallowly will dry out in the autumn, when they should be making their flower buds for the spring, and will come up 'blind', that is, without flowers. There are few more pointless sights than a sea of daffodil leaves. So daffodil bulbs, which are three inches long, must be planted with their roots embedded in grit nine inches below ground.

Every inch had to be fought for. Never, ever, will I advise someone to dig a runner-bean trench, plant bulbs in grass, or double-dig a new border without trying to imagine what those seemingly simple acts require for anyone who gardens on a heavy soil.

All the livelong afternoon I glanced restlessly skywards for rain that would respectably end that dreary, painfully slow, arm- and back-aching occupation. At long last the area dug exceeded that still to be done, and those daffodils danced once more in my head. But I had discovered something disturbing, from which I had hitherto been shielded by my previous, light soil. The most necessary attribute for a gardener now seems to be not 'green fingers', or botanical expertise, or even money – but courage. And that had almost failed me.

Season of Discontent

If there is a better instance of the malevolence of inanimate objects than that exhibited by plastic garden netting, I should like to meet it. Or, more truthfully, I should *not* like to meet it. This green, polypropylene, 'flexible', open-weave netting has been in place to support my sweet peas since the day in early June when I had the fiddling task of taking it out of its box, unwrapping it, threading it at intervals with suitable bamboo canes, and tying it and them together with string. All this to ensure that it stayed upright and taut and did not slide gently down, like a drunk subsiding to the barroom floor.

Now that the sweet peas are over, the structure must be dismantled, a job which fills me with even less enthusiasm. I have now to remove each stem which has insinuated its way among the strands of netting. If I wait until the stems have dried to a crisp before pulling them off, which makes the job much easier, I must first find somewhere dry to store the netting.

Even when all the stems are pulled away I shall not be able to roll the netting up sufficiently neatly for it to go back in its original box, even had I had the sense to keep it. So, instead, I will tie string round it to make a bulky netting sausage and hang it from a nail in the shed, at a height at which it will smother or buffet me every time I go to fetch a tool.

The Murphy's Law season is upon me: by the time the netting has finally been folded and put away, it will be time to think about greasebands. These are not, as you might think, leather-jacketed rock groups, but the sticky bands put round fruit trees to catch flightless female winter moths as they climb up the tree trunks to lay their eggs in crevices. The greasebands come on a long roll, with the sticky part to the inside. As you pull away the outer covering, your hand inevitably comes into contact with the inside. These bands have, for some reason, been devised to stick to things far more substantial than wingless winter moths. They and I usually become as inseparable as newly-weds.

It is also time to plant spring-flowering bulbs. This occupation convinces me that animate objects also have an animus against gardeners. It is a reasonable rule of thumb that, wherever you want to plant a bulb, there will already be one. This happens as a matter of course if you are planting bulbs for spring bedding display in a border, but is not unknown in the wide open spaces of lawn or orchard. And, on the subject of spring bedding, have you noticed that however careful you are to plant every single wallflower that you have placed on the soil, there will always be two that you notice a few days later, lying shrivelled and beyond salvation?

This is also the season when gardeners find themselves spending more time in the greenhouse, and here too you are likely to encounter the implacable enmity of Natural Law. The erection of bubble polythene insulation is a fertile ground for frustration, and there is no chance of washing down the house without getting disinfectant-laden water down your neck and arm and in your Wellington boot. The last tomatoes in the grow-bag will defeat any effort you make to keep them well-watered, for the packed roots will ensure that all the water slides over the edge of the bag.

However, malevolent objects do not always have their own way. At least our hosepipes no longer kink. I remember when as a child I was asked to wash the car, how the hosepipe refused to unravel and, even more, to ravel up again. Even when the damn' thing was pulled right out, it had a way of catching on every stone or step. The modern threaded hosepipe, flexible and amenable in a way its forerunner never was, and with a custom-made holder on wheels onto which it can be reeled, constitutes genuine progress.

But there is still scope for inventing devices which will circumvent Murphy's Law: a machine for spreading lawn seed evenly; a blunt-tined fork for digging up potatoes, such as used to exist but which can no longer be found; genetically-engineered wasp-repellent plums; and, most useful of all, sweet pea supports consisting of netting already threaded and attached to sturdy canes. Now, they really would show who was the boss in the garden.

Stick Around   15 November 1997

How important it is for humanity always to make a virtue out of necessity. Now that affordable, highly-skilled professional gardening help is as rare as the black-tailed godwit, we are ingenious in finding reasons for not doing things in the garden. 'Labour-saving' has been elevated into a positive virtue, rather than accepted as a desperate response to unfriendly circumstances. There may not yet be too many people who assert that every plant must be allowed to develop in its own particular way, blossoming and flourishing, withering and perishing, without much help or hindrance from judgemental, control-freak gardeners, but there is an ever-increasing band of people who have decided that gardening is what you do only between Easter Bank Holiday Monday and the Saturday that the clocks go back. Whenever they may be.

The rest of us have quietly abandoned many jobs that we once did, dismissing much autumn and winter work in particular as out-of-date. As most plants are sold in containers these days, it stands to reason that only old fuddyduddies will bother to seek out bare-rooted hardy trees, shrubs and roses to plant in the dormant season. What is the point of raking fallen leaves if you can leave them to lie on the soil as a mulch or, more likely, to blow around the garden? I should be surprised if many people think to sweep worm casts off lawns these days, and almost nobody forces rhubarb in the greenhouse. To be fair, even I have come round to the view that, on certain soils, winter digging – that great standby for industrious gardeners in the past – is just so much waste of time.

The other great standby, the cutting-back of hardy perennials in the autumn, has lost popularity since we were taught to admire the quiet beauty of sere ornamental grasses and flower seedheads when rimed with frost on sunny winter mornings. We have seen, in illustrated books, what charming effects can be achieved by leaving well alone. The only problem is that we will be lucky if this happens more than half a dozen times in a season, and we probably won't be awake when it does.

I am not suggesting that gardeners take up once more the banner of barren tidy-mindedness and clear away everything that is fading in the borders, leaving nothing but brown earth and the crowns of plants, especially as garden birds undoubtedly benefit from a few seedheads left in winter. However, it is easy to forget that the garden is home to some indomitable and selfish egos (and that's without counting the members of your family), and the gardener, as referee, is at her most powerful (or least powerless) in the autumn, when leafy passions are spent. No border works well if left alone for more than two or three seasons; the dynamic of growth and development is too relentless for that. So there is still some truth in the old head-gardener maxim 'Hurry in autumn, tarry in spring.' Those who do absolutely nothing in autumn run the risk that when spring arrives, mayhem will break out, and far too much time will be spent in tedious fire-fighting.

The Indian summer of the autumn just past has been more than usually pleasurable for anyone who has not yet lost interest in controlling ferocious seeders and colonisers. The soil was moist, not wet, and the days, after sharp frosts, were balmy. Dividing and throwing out too-vigorous herbaceous perennials, therefore, was not a fight with muddy hands and slippery spades but could be done decorously under cloudless skies. Although the old injunction to dig up and divide all perennials every three years has rightly been abandoned, hearty ones

like members of the daisy family do need curbing and renew-
ing from time to time. It may help to know that root growth,
which precedes stem growth after a plant is transplanted, continues into December but is slow to start in spring. Root cuttings, a thoroughly unfashionable but easy way of propagating
Oriental poppies and phlox, can be taken now.

As the legitimate occupiers of the border shrink and falter,
perennial weeds which were born to blush unseen are forced
into the light of common day. This is one moment to hammer
bindweed, in particular. Although its leaves go yellow and fall
in October, the stringy lianes of stems remain, still attached to
the ruthlessly imperialist white roots below. In autumn there is
time and space to burrow for them.

There is another, rather more positive, reason for spending
the autumn and early winter in the borders. Rearing plants,
caring for them as they burgeon and flower, then ignoring
them until you come across their frosted, dried-stick remains
next Easter, is like rearing children to adolescence and then
abandoning them, later failing to recognise them in the street.
Gardening is only really fun, and successful, if you know your
plants well, and for that you need to be around at the end just
as much as the beginning.

6

Theoretical Gardening

What are Old Crocks? *5 August 1995*

Thank goodness I do not write about bridge. Or chess. 'He won East's heart return, drew the remaining trumps and ruffed both his losing hearts in dummy . . . Black suffers from multiple fracturing of his pawns.'

My colleagues have no space to make themselves intelligible to anyone who has never played any bridge or chess. I, on the other hand, can depend on most people having picked up some gardening jargon along the way, however hard they have tried to avoid it.

Nevertheless, although my sentences can be understood in general terms by someone with no practical gardening experience, there are a number of horticultural mysteries wrapped up in what appear to be meaningful words. I am, therefore, taking this opportunity to clear up a few common confusions.

'Humus' is a Greek-style *hors d'oeuvre* which can be bought cheaply in quantity from Waitrose on a Saturday afternoon when about to pass its sell-by date and used to dig in to heavy clay soils. Its unique flocculating properties make the soil particles aggregate, thus opening up air holes and ensuring a better-drained soil.

'Old crocks' are people you employ, at a ruinously expensive hourly rate, to come in one day a week to suck their teeth at the pea midge in the vegetable garden and refuse to climb ladders because they suffer from verdigris.

A 'self-clinging climber' is one that enfolds you in a passionate embrace as you come out of the front door. A good example is Virginia Creeper.

A 'rambler' is a 'self-incompatible' 'maiden' who insists that there is an ancient right-of-way through your rose garden. Anyone who experiences difficulties would be wise to erect a 'cordon' around the garden's perimeter. A 'panning' can be helpful in eradicating this pest but 'whips' should only be used as a last resort.

'Bastard-trenching' is another expression for double-digging, an activity favoured by professional gardeners, who are the only people prepared to put themselves through this torture; it is so-called because it is a right bastard to do properly.

'Scarifying' is the process whereby a propagator intimidates seeds into sprouting by threatening them with a metal file.

'Leaf mould' is a fungus which grows on leaves, causing them to fall prematurely.

A 'floating cloche' is a lightweight polypropylene sheet laid on a garden pond to protect water plants from frost. These water plants are known as 'marginals', because it is a toss-up whether you can be bothered to grow such invasive and dubiously attractive plants.

'Thatch' is the dead grass which lies on the top of the lawn and, when raked up, can be used for re-roofing the rustic summer-house.

'Offset' is short for the expression 'offset on the farm'; it is a widely accepted abbreviation used by those wishing to hide the gardener's wages and petrol from the Inland Revenue.

'Organic' gardeners are made up of compounds that contain carbon. When 'organic' gardeners die, their remains can be used as a 'mulch' or 'compost', for retaining moisture in the soil and keeping weeds under control. The rest of us are 'inorganic', which means that we degrade far more slowly and, what is more, we need expensive and non-renewable fossil fuels for our manufacture.

A 'spade' is (a) one of the two black marks by which one of the four suits in a pack of cards is distinguished; and (b) a bloody shovel.

Now that these words have been explained to you, I hope that gardening will lose something of its infuriating mystery. Knowledge of this kind should ensure that you are not intimidated by any other arcane term that a gardener may use to guard his or her preserve. What is more, you should never forget that even professionals do not know it all. There is one mystery, for example, which I shall never fathom – namely, why the propagation of herbaceous plants is known as 'division', not 'multiplication'. No, please don't take the trouble to write.

Rosa *by Any Other Name*   *28 March 1987*

To the outside world, the line between the botanist and the horticulturist is rather fuzzily drawn, seeming more a matter of nomenclature than substance. Botanists are concerned with the classification and naming of plants, whereas horticulturists are cultivators of plants and care little, if at all, for the way their charges come to be named. Botanists are sometimes also gardeners, almost as interested in living flowers as in dried herbarium specimens; gardeners are hardly ever also botanists.

The botanist seeks diligently after a realisable truth: that is, the correct name for a plant. Since the mid eighteenth century, when Linnaeus classified all known plants and animals using the binomial system, names have consisted of a generic term (akin to a surname but coming first) followed by a specific epithet (like a forename but put second). The classification is based on the sexual characteristics of the plant. In the twentieth century the introduction of the International Code of Botanical Nomenclature standardised and simplified the naming of plants still further, but enshrined the apparent absurdity – to gardeners, at least – of the principle of 'priority'. In the process of establishing 'priority' (the earliest record of the naming of a particular species) or in taking account of advances in taxonomy, the botanist may have to correct the imprecision or ignorance of earlier botanists by making name changes. Renaming a plant may involve putting it in a different genus or even family.

The professional horticulturist, on the other hand, is not the slightest bit interested in whether de Candolle named this plant 'x', or whether Linnaeus got there first with 'y', or whether both men were even referring to the same plant. Nor does he understand why some botanists, the 'lumpers', put species together under one name and others, the 'splitters', are keener on dividing them up. The gardener craves certainty, as the botanist does, but for her that means preserving the status quo. Nothing infuriates her more than to have grown for many years the South African half-hardy, blue-flowered *Agapanthus umbellatus*, only to find that when she orders an interesting 'new' species called *A. africanus* from a catalogue she receives yet more of the plant which she already possesses. What does she care that the new name is earlier and botanically more correct? Botanists renaming plants have led to gardeners renaming botanists.

Just as cross and loud in complaint is the average amateur gardener whose knowledge of the classics is a little shaky. Her irritation is not so much with name changes as with the names themselves. She thinks she would much rather that *Agapanthus africanus* could be found in catalogues as 'African lily', and wails that it is a great shame that people cannot call plants by their proper names. She forgets, of course, how many botanical Latin names, like *Clematis*, she uses as a matter of course.

There is reason enough to retain in use the common names of native British flowers, and most gardeners, if they are not too pedantic, would prefer heart's-ease to *Viola tricolor* and primrose to *Primula vulgaris*, but the eight regional names for the marsh marigold (I mean marsh buttercup, sorry, molly-blobs, no, kingcup) make one fall back with gratitude on the universally comprehensible *Caltha palustris*.

I would readily concede, therefore, that this is a matter about which we are none of us entirely consistent, particularly because even the Latin or Greek can sometimes be misleading. For example, *Rosa centifolia* has a 'hundred' (i.e., many) petals, not a hundred leaves, as the name would suggest.

The controversy surrounding this subject is anything but new. More than a hundred years ago, William Robinson, a horticultural polemicist of genius much influenced by John Ruskin's dottier outpourings, translated many of the names of plants he used in *The Wild Garden* into English, in a laudable but misguided attempt to make life easier for the ordinary gardener. In the process he allowed himself to fall into the absurdity of spurious anglicisations such as 'rockfoil' for *Saxifraga* and 'hair bell virgin's bower' for *Clematis campaniflora*. I should feel a reticence about asking for the latter in my local nursery. Some plants, such as roses and daffodils, have been around so long in so many cultivated forms that it is neither necessary nor even desirable to mention their Latin names, but is anybody the wiser if *Agapanthus* is called by its English name? 'African lily' tells us almost nothing useful about the plant; indeed, as the plant is not a lily the name is positively misleading.

I am no botanist (although so vague is the world's perception of the difference between botany and horticulture that I am frequently introduced as one), but in this matter I side with them against my own kind. I have worked with very able young gardeners whose written English was pretty shaky but who had no difficulty, because they had no choice, in learning to say and spell *Metasequoia glyptostroboides* or *Gentiana przewalskii*.

It seems to me that as botanists do not change names merely to infuriate gardeners, although it must be tempting sometimes to tease such a captious lot, and as most names are constant, we gardeners have little of which we can legitimately complain. The botanist Linnaeus's use of the binomial system provided a nomenclature which could, and can, be understood all over the world, and which succeeded in simplifying names which were frequently ten words long. We gardeners should be grateful for small mercies.

One of the perks of being a gardening writer is that you are sent a great many gardening books in the course of a year. Admittedly they are of variable quality, but there are usually more than enough which deserve an airing. Often the only way of dealing fairly with them, as there are so many, is to write about them in a 'round-up' review before Christmas. The following is such a review, which I fancy has not lost its relevance in the intervening years. Indeed, some years after it appeared, a reader wrote to express his disappointment at not being able to find The Ducal Gardens of Basilicata *anywhere, and wondered if I might be able to help.*

Garden Leaves   *23/30 December 1989*

A more than usually bumper crop of gardening books has rained down around me this year. I was particularly struck by *The Hanoverian Country Gentleman's Kitchen Garden Diary and*

Address Book, which weighs two and three-quarter pounds. This startlingly original and lavishly illustrated work was unexpectedly found under some old rubbish in a corner of the imagination of a book packager, where it had lain undiscovered for nearly three minutes. There is much useful advice on making hotbeds out of pigeon guano, plashing medlars, and what to do about those unsightly split ends when your topiary starts to grow out. The book includes free 'Georgian' seed packets and a pop-up trug.

Dora Appletree's new book of gardening reminiscences, *My Garden at Whitsun*, is sure to delight her numerous fans and make her many more. With wise words to say on such diverse subjects as *Syringa vulgaris* and the common lilac, a hundred and one uses for dried hydrangea flower heads, and how to make your own larch-pole arbour, this is surely a fitting sequel to *Spring in my Garden*, *The Summer Garden*, *Gardening in Autumn* and *Wintry Gardens*. This lavishly illustrated book contains many photographs which have become old favourites. We can only hold our breaths and wait, impatiently, for *My May Day Bank Holiday Garden*.

I predict that no book will take the gardening public so much by storm this Christmas as *Some Favourite Gesneriads of Central Africa* by Dr Phil Ode, a heartwarming account of those cuddly little saintpaulias known to their devotees in bathrooms up and down the country as 'African violets'. But behind the fun this book has a serious purpose, exposing the connection between their chromosome count and a regrettable weakness for polyploidy. Lavishly illustrated with pen-and-ink drawings of the parts of the flower and hilarious monochrome photographs of Dr Ode's party of botanists pitching camp in the Zairean jungle, this book is a must for any gardener's unwearoutable sock or stocking.

I wish I could say the same for *Designing and Planting Your Garden* by Vi Burnham. This is a very disappointing book with only limited appeal to the gardening public. It sets out to advise the gardener on how to make his or her garden a place of tranquil beauty and harmonious colour for twelve months of the year, using plans, three-dimensional drawings and colour photographs. Looking out of place on even the smallest coffee table, this book could only be of any use to those 'odd-balls' who spend more time working in their gardens than talking about them.

At last! A Japanese gardening book which loses almost nothing in translation. Taiotoshi Kosotogari's *Cherry, Stone and Water* looks set to blow a gale of fresh air through those fuddy-duddy gardens locked in a Jekyllesque time-warp. We can only hope that space will now be found at Sissinghurst or Hidcote for areas of raked gravel, bonsai maples and a few impeccably chosen and immaculately placed rocks. What a strange irony that the photographs which lavishly illustrate this book were all taken with a Hasselblad!

For light relief, I turned to *The Ducal Gardens of Basilicata* by the Principessa Rigovernatura-Sporca (with Sandy Poole). This fascinating and absorbing book bears ample witness to the wealth of talent there is among the ducal families of this favoured spot, the 'instep of Italy'. How happy and privileged one feels to be beckoned into their ancestral demesnes, crowded with Mediterranean cypresses (sadly, in places, suffering from the wretched cypress disease which has swept Italy in recent years), box hedges and fountains. It is a great shame that when the photographs were taken for the lavish illustrations, an unusual summer drought should have dictated that the fountains were not playing nor the grass watered. Nevertheless, the gardens are enchantment itself and deserve to be more widely known. If I have a quibble, it is that although the nicely-spoken girl from the PR firm who sent me the book assured me that the Principessa was 'really closely involved and whatever' in the preparation of the book, it is not always easy to tell, in relation to any particular portion of the text, whether the Principessa or Miss Poole should take the credit.

Evenings of Wine and Roses   16 July 1988

There is nothing, nothing, quite so nice as sitting about on a warm summer's evening, glass of wine in hand, within smelling distance of 'Madame Lauriol de Barny'. This rose has a rich, fruity scent usually called, for want of anything better, evocative. For me, it recalls memories of nothing so much as sitting about on a warm summer's evening, glass of wine in hand . . .

We would not need to fall back on vague words like 'evocative' if there were as developed a vocabulary to describe scent as there is for colour. Perhaps we could learn something from the wine buff who has sweated to find ways of describing the indescribable, if not the unspeakable; after all, a good 'nose' for wine and one for flower scent often go together.

Take my husband, for example. His knowledge of gardening is rudimentary and reluctant, but his habit of defining the tastes of any number of inexpensive wines from a variety of unlikely countries makes his an invaluable 'nose' in the garden. I would do well to copy his way of scribbling down a few helpful notes: '*Hurdia ashdownensia* 1983 Hattersley Garden Centre (£6.95) planted next to Lawson cypress. Tarmac, softening to rhubarb fool and brass screws, with hint of hamster sawdust – buy more!'

I find his comments particularly helpful, for my own sense of smell is not much more than adequate. A youth misspent smoking cheap cigarettes, together with an enduring susceptibility to hay fever which lasts far longer than the pollen season, have done nothing to improve what was only ever average. That is my excuse, although hardly a convincing explanation. After all, the botanist and gardener E.A. Bowles (1865–1954) was so disabled by hay fever that he took to spending June in the European Alps to avoid the flowering grasses at home; yet, by popular consent, he had a very fine 'nose'. I have discovered for myself that *Cytisus battandieri* smells of fruit salad, but it needs a Bowles to add 'with a dash of maraschino or kirsch'. I envy my husband his hooter – not its physiognomy, it being even larger than my own, but its sensitivity. It was he who exploded, to my satisfaction at least, the widely-held myth that the flowers of *Iris graminea* smell of ripe plums when in fact they smell of apricots. I have tried this idea on a number of people since, all of whom have been forced to agree. I find that smelling the rose 'Madame Alfred Carrière' has an added savour since he identified the scent as that of tinned lychees.

Flower scents arise from the presence of essential oils, called attars, which are composed of a mix of chemical compounds. Citral, for example, is the dominant ester in lemon flowers, benzyl acetate in jasmine. Scents are classified into several major groups. We tend to respond favourably to those in the spicily sweet Aromatic group, such as carnations, heliotrope and witch hazel, which are pollinated mainly by butterflies, while positively avoiding those in the Indoloid group, which smell of rotten meat in order to attract a variety of flies.

After all, although it would be agreeable to believe otherwise, flowers do not have scent for our sensual benefit. It is no coincidence that the paler the colour of flower, the more likely it is to have a strong scent. Few deep-blue flowers, for example, are strongly scented, whereas the pale mauve-pink night-scented stock, *Matthiola bicornis*, can waft the smell of aniseed for many yards.

Comparisons made to identify scents have their danger, of course, just as they do when tasting wines. If you pigeon-hole Cabernet Sauvignon as tasting of blackcurrants, you may deny all Cabernet Sauvignon varietals the possibility of being perceived as more complicated than that. The same is true of flower scents. Mahonia flowers, for example, are often said to smell of lily-of-the-valley, when actually they smell of mahonia. This genus may have acquired its olfactory reputation because lily-of-the-valley has been the scent of certain soaps for generations and so is well-known, whereas the mahonia has not, and is not. (I certainly knew the smell of gardenia from the soap long before I ever smelled the flowers.) But using the example of a soap will not wash.

I blinded myself to the real scent of *Iris graminea* by detecting the smell of ripe plums because I had been told it was there – although I cannot exclude the possibility that it is only the clone which I have which smells of apricot, just as one of my two 'Zéphirine Drouhin' roses is much more fragrant than the other. Aspect and, particularly, climate are also factors affecting the strength of scent. Lavender is supposed to smell more strongly in Britain, where conditions are cooler, than in its native Mediterranean region.

Appreciation of scent is so personal that reading other people's opinions on the subject is often next to useless, and can be actively misleading; learning to identify the essence of fragrances can only be done by practice and by attempting to put into words their complexity. It need not only be children who can enjoy guessing games which one might, for want of a better phrase, call 'blind smellings'. And, though this is a suggestion unlikely to find favour among the killjoys of the British Medical Association, it would obviously do us all good to drink more wine.

All Squared Up   28 *June 1986*

Is it, I wonder, a fascist tendency, a fatal weakness for the smack of strong government, perhaps, which underlies the enthusiasm with which I cultivate my vegetable garden? Why else should I so admire the neatness and orderliness inherent in growing vegetables when, in the process, the exercise of choice or imagination is largely suppressed and individuality crushed? No free-thinking notions of beauty or spiritual enrichment can apply here, after all; it is a far cry from the anarchic freedom of my flower garden. I would not tease myself with these questions were it not that I find myself more often spending an hour on a summer's evening, brain in neutral, thinning carrots than struggling with decisions as difficult as how best to train a wayward clematis. I feel a little guilty about preferring vegetable gardening, and wish to punish myself for enjoying it.

Fortunately for me, no such disturbing anxieties appear to ruffle Mr P's placid calm. Every Wednesday morning he comes, in his words, 'to sort me out', and I must yield to him the privilege of overseeing the kitchen garden. In the flower garden he is more timid, for he is asked to hoe carefully around what he feels sure must be weeds and is rarely allowed to give shrubs a good trim; to him the rules seem haywire and the rationale obscure. Though thorough and conscientious, he appears to long for the security to be found beyond the vegetable garden wall. So, sometimes, do I.

Mr P. fits every patronising garden owner's vision of the sturdy, skilful pensioner ('we have this simply *marvellous* little man'). I suspect strongly that he would not care to be patronised, however, and he certainly enjoys to the full the irony that he is a better vegetable gardener than I am. This is a fact which I willingly concede. I live in hope that when I am seventy I shall be able to sow a ruler-straight drill without a line and know to the day when the blackfly will begin to descend on the broad beans.

His pleasure consists in getting me 'all squared up', probably believing that when his back is turned I shall go round in circles. He is partly right. I am certainly sometimes borne down by the weeds, but he neither flinches from nettles nor shrinks from ground elder. In his turn, he will admit that I am *sometimes* right, certainly about new varieties, and he is just polite enough to refrain from asking me why I bother with small tomatoes like 'Gardener's Delight' when we all know what a good cropper 'Moneymaker' is. I have always enjoyed watching other gardeners garden, and I have learned a great deal from observing his unhurried ways. He brings me straw-berries on my birthday, the date of which he elicited from his wife, Madam President of the WI, and he brings the children presents each Christmas Eve. We are friends, and I shall be very sorry when he finally hangs up his hoe.

Even if his sensible pride did not reassure me, I should not have to search far for a reason for pursuing such a blatantly utilitarian form of gardening. After all, the vegetable garden has considerable, if restrained, potential for beauty. With brick and stone paths, parsley pots and tubs, sweet peas growing up the fruit cage netting, variegated cabbages and coloured-foliage sages, it is easy enough to make a pretty effect. Those who have seen the ornamental *potager* at Villandry, near Tours, will know what is possible. If single-flowered French mari-golds are planted to discourage the whitefly on tomatoes and nasturtiums to see off aphids on the lettuce, some floral element, satisfactorily tamed and arranged, may be safely and happily admitted.

I wish I could say that the number of committed vegetable gardeners is growing,* but the widespread evidence of aban-doned allotments is too obvious to be ignored. This is partly because, with so many other competing distractions, the work seems too labour-intensive; but it is also due to the ubiquity of freezers, which make the growing of quantities of space- and time-consuming winter brassicas redundant.

It is a pity, because never have there been so many interest-ing vegetables to grow. The list is long – golden zucchini, 'Pink Fir Apple' potatoes, striped 'tiger' tomatoes, vegetable oyster, asparagus peas, Hamburg parsley; there seems no end to the Japanese onions, Chinese greens and American corn which

---

*In the past fifteen years this trend has gradually been reversed and, though allotments are no more popular, seedsmen now report sales of vegetable seeds outstripping flowers. This is in great part due to concerns about healthy eating, as well as to the increasing availability of the kinds of inter-esting vegetables mentioned in the next paragraph.

somehow, despite the limitations of day-length and climate, we are capable of growing.

Mr P. and I need not concern ourselves with any airy-fairy idea about retreat from an unsatisfactory and chaotic world; in any event, we have not the time to do so, with the runner bean supports to erect and the cauliflowers to transplant. For our own benefit we may continue to use the cultivation of vegetables partly as a welcome escape from the difficult decisions inherent in the making of a flower garden, but most of all we shall enjoy the pleasure which results from being 'all squared up'.

Tommy This and Tommy That 16 February 1991

It is hard to know quite why the word 'amateur' should so often be used pejoratively. After all, the beauty of being an amateur is that it frees you from all the constraints inevitably imposed by money. I grant this is a contention unlikely to find much favour with the England XV* but in gardening, at least, the ability to do what you like usually outweighs any financial loss.

Nowhere is the gap between commercial limitations and liberated amateurism more evident than in the growing of vegetables, particularly tomatoes. Commercial tomato growing is a cheerless, highly technical, automated affair and, in this climate, not without its uncertainties. The varieties grown for the market must be high-yielding, thick-skinned and uniform in size, which is why tasteless 'Abunda' is so abundant. They must also be picked before they are quite ripe. The private gardener, on the other hand, is free to grow small but ripe crops of delicious, thin-skinned, variously-sized tomatoes, like 'Sweet

---

*This was written at a time when there was a lot of debate about amateur rugby union players turning professional.

100' and 'Gardener's Delight', the kind which would not keep a business afloat for a fortnight.

Tomatoes are as popular as any vegetable (or berry, if you wish to be pedantic) and more so than most. They are grown by gardeners who would not soil their hands with the commonplace cabbage or *passé* parsnip. This is because tomatoes are never dirt cheap and are both good fun and relatively easy to cultivate. Curiously, although introduced to Europe from South America in 1595, they have only been widely grown for culinary use since the mid nineteenth century. Before that, they were considered decorative but poisonous, like vamps and vipers, and given a wide berth. Except in Italy, where the cooks were adventurous.

They can, within reason, be grown anywhere, provided they are not planted outside until after the frosts. My brother, who hates gardening but loves tomatoes, used to grow plants of the Continental variety 'Marmande' and the stripy 'Tigerella' in pots on the sill of a tall window in Garrick Street. They blocked the light and attracted whitefly, goodness knows from where, but he nevertheless harvested a reasonable crop over a long period. Late in the season he would wrap the fruits in newspaper and ripen them in a drawer of his desk, so that there would be tomatoes on the table at the office Christmas lunch.

It is ironic that the amateur's freedom to grow tomatoes in almost any way except actually upside down – as bushes or as cordons in the open, under cloches, in cold or heated greenhouses – is due in great part to the research done to improve commercial yields. In particular, the breeding of tomato varieties naturally resistant to diseases and disorders, and even a resistant rootstock (KNVF) on to which a root-disease-prone variety may be grafted, have been of benefit to the amateur.

So too has the development of techniques such as 'ring', strawbale and growing-bag culture. It is not too fanciful to foresee a time when even advanced commercial practices such as 'nutrient film technique', where tomatoes are grown in nutrient-rich running water, or a variant of it which uses mineral rockwool as a planting medium, will be adopted by serious gardeners searching for an acceptable alternative to peat.

The amateur can use commercial technology yet be free of imperatives concerned with plant density, yield and size. She need not buy expensive F 1 hybrid seed which produces vigorous and uniform plants, nor need she grow tomatoes indoors. What is more, as a beginner, uncertain of the pitfalls, she can buy young plants of an outdoor 'bush' variety at the garden centre, plant them out in a sunny place and, apart from copious watering and some liquid feeding, leave them alone to flourish. These 'bush' tomatoes need not be staked or have their sideshoots removed, thus avoiding both work and the sickly nightshade smell of crushed tomato leaf which lingers about one like the odour of sanctity. I take pleasure in the fact that one of the best of the foolproof 'bush' varieties is called – no doubt with damning intent – 'The Amateur'.

Monster Girths   15 October 1994

There are several profound differences between men and women, as we all know. Women can do crochet work but cannot cut a loaf of bread straight. Women can make a mental shopping list while reading a bedtime story, but cannot listen to the football results all the way through. Women will happily grow 'baby' vegetables for the kitchen, but they will not grow them to gargantuan size.

This was plain last week at the *Garden News* Giants Championship at Baytrees garden centre near Spalding, where the hundred competitors were almost all male. You can see why, easily enough. Although the vegetables must be in sound condition to qualify for a prize, what they look like is quite beside the point; the woody knobbliness and inutility of many of these vegetables is unlikely to appeal to women. Although surprisingly many (the giant pumpkins, marrows, mooli radishes, onions, tomatoes and cabbages) are eatable, they are just too big to be anything but a flaming nuisance in the kitchen.

So why do they do it? Some admitted to growing giants because their too-easy success at 'quality', as opposed to 'size', shows had stripped these of excitement. Although they may still chance their arm on getting the perfect five 8-oz. onions onto the show bench, what really gives them a thrill is to hear the words 'New World Record'.

Moreover, unlike most conventional produce shows, this one attracts enough sponsorship to mean decent prizes. The grower of the heaviest pumpkin wins £250, with a second prize of £75, third of £25, and fifteen more prizes of £10. As a fair few of the serious competitors are unemployed or retired, this comes in, as they say, very handy.

It is impossible to imagine how anyone with much else to do would have a chance here, anyway; these vegetables need almost constant care and attention to achieve maximum size and weight. The moment for picking a heavy tomato, for example, is gauged when its girth no longer grows by one and a quarter inches a day.

These men's knowledge of plant physiology is impressive. As money is often tight, they depend heavily on manures and home-made concoctions to feed the vegetables, preferring to use any spare cash for polythene tunnels, grow-lights for onions, and plastic guttering in which to grow entries for the longest parsnip or carrot classes. These two classes are a bit of a chizz, incidentally: most of the vegetable's length comes from strand-like fibrous, rather than 'edible', roots. Generally speaking, there are no magic formulae; most amateur gardeners could not grow monsters if they tried, because their soil is not cultivated and fertilised to anywhere near its full potential – and, of course, they are not prepared to leave the vegetables in the ground all summer.

Stories have always circulated about how some growers 'bend the rules'; for example, by inflating marrows using a sugar-and-water solution drawn up by a cotton-wool wick attached to the stalk. I was told in no uncertain terms that this is not done, mainly because it does not work, but the chief safeguard against cheating is that most judges have also, in the past, been exhibitors.

This was not a year for breaking world records. A wet, cold spring followed by high temperatures in July had inhibited growth at crucial times. But Ian Neale of Newport, a market gardener, managed it with the heaviest beetroot. A 40 lb. 8 oz. beetroot is scarcely recognisable as such, being almost square and as warty as an old toad. It is a good thing that it is uneatable because it must be replanted after the competition, to flower and set seed; this is then sown, or given away to best mates.

These growers set great store by the strain of seed. Genealogy matters as much to them as to a duke's second cousin. Bernard Lavery, who holds the world record for a pumpkin (710 lb.) and who acted as impresario and compère to this entertainment, has turned his capabilities to commercial advantage, founding a seed club and selling seed of giants for £2 a packet. He is keen to emphasise the fun of it all. Pumpkins certainly make me laugh because they resemble recently vacated bean bags, but much of the rest is too grotesque to be funny.

There is certainly nothing laughable about the growers. Contrary to popular myth, they are not wild-eyed obsessives. Richard Hope of Wigan, for example, who produced both a 74-lb. marrow and 15-lb. cucumber, is a smiling, softly-spoken, craggy version of Will Carling who looks very much at peace with himself. Easy to spot in the crowd because of their crinkly-eyed sunburnt faces and home-knitted jerseys, the showmen could not be seen so much as exchanging a cross word, let alone indulging in an unseemly row. They may be competitors, but they know they are part of the freemasonry of giant veg. growers, whom the world – or that part of it which is female, at least – simply does not understand.

Health Warning   *19 February 2000*

Just when you thought it was safe to go back in the garden after the winter, I am here to tell you that it is not. In fact, you would be better off doing your weeding on a bypass roundabout than in the cosy, familiar privacy of your back garden. The garden where you can give expression to your creative instincts or settle your jangling nerves with honest toil is a complex of traps and pitfalls of which a jungle would be proud. According to the Consumer Affairs Directorate of the Department of Trade and Industry, 20 per cent of non-fatal accidents in the home relate to gardening. In 1996, the last year for which the directorate can give me figures, there were 464,000 injuries.

At times we contribute all too blithely to these statistics. We wear treadless rubber boots, run down frosty steps, prune whippy climbing roses without eye protection, stretch too far on ladders, leave garden tools lying about in long grass, forget to plug in the Residual Current Device when switching on the electric hover mower, and would not dream of doing warming-up exercises before beginning on the winter digging. No wonder the accidents pile up.

The situation is even worse than that, however. Accidents can happen anywhere, but there are gardening activities which are inherently deleterious to health. For example, Professor Mike Griffin of the Institute of Sound and Vibration Research was reported recently in *Gardening Which?* as saying that anyone who uses power-tools such as mowers and chainsaws regularly or over long periods risks Hand–Arm Vibration Syndrome, the symptoms of which include tingling, numbness, and cold and white fingers.

Now, we gardeners know a thing or two about syndromes. One to which we are particularly prone is Finger Cutting Syndrome, a common occurrence if care is not taken with a new pair of secateurs, while Eye Poking Syndrome affects all who don't watch where they are going when pushing through the shrubbery or weeding between staked plants. My particular bugbear is Expletive Explosion Syndrome, which struck me particularly badly the day I put a garden fork through my foot.

Power-tools are pretty bad for syndromes, I can tell you. Although they have been improved in recent years by the adoption of safety features such as plastic blades on small rotary mowers, automatic cut-outs, and anti-vibration systems on some chainsaws, nevertheless we can risk not only Hand–Arm Vibration Syndrome but also what Professor Griffin would probably call Ear Loud-Noise Syndrome, the symptoms of which are tinnitus and hearing loss. Petrol-driven garden machinery can be extremely noisy, although to be fair to manufacturers there have been attempts to improve matters on this front too. Nevertheless, particularly with large and old machines, there is still a problem. The generation presently most furiously engaged in gardening is the one which assaulted its ears at Pink Floyd concerts in the 1960s. If its hearing was not damaged then, it will probably not have survived intact decades of once-weekly mowing expeditions and three-monthly shredder or strimmer orgies.

I mention this one in particular because, unlike many gardening accidents which, though they may be foreseen, are not always easily prevented, hearing damage is perfectly avoidable. All you need is a pair of ear-muffs, of the type which are sold to game- and clay-pigeon shots. These muffs cover the entire ear and, though strikingly unglamorous, make mowing in the privacy of your own garden a great deal more pleasant.

If all this sounds dreary and earnest, I have to say that there is a tangential advantage to these muffs which I commend to all those people with reluctant gardening spouses. It is possible to wear the earpieces for a portable radio or disc player underneath them. My husband, for example, favours a small long-wave radio, with a clip to attach it to pocket or belt, which you can buy for a tenner at cricket grounds for listening to *Test Match Special*. Indeed, the fact that the cricket season more or less coincides with the grass growing is a great boon to me. Cricket being the stately, paced game it is, there is no problem in getting him in for lunch – which is lucky, for he could certainly not hear my shouts. The only problem is that, according to DTI statistics, he is even more likely to suffer some kind of harm once he gets back inside the house. Help!

Non-stick Wicket   9 August 1986

Cricketers seem addicted to gardening. Sometimes scarcely an over seems to pass without the batsman raising a hand, as the bowler starts his run-up, in order to advance down the wicket to peer suspiciously at the surface and prod it repeatedly with his bat. But is there really much to connect the making of a cricket square with 'gardening' in its more conventional sense?

On the face of it, perhaps not. The gardener and groundsman certainly feel they have little in common. Ron Allsopp, the head groundsman at Trent Bridge, whose expertise is matched only by his enthusiasm and willingness to share his knowledge, confessed to me that he cannot stand gardening, and I dozed fitfully through my groundsmanship lectures at Kew, as is demonstrated by the indecipherable squiggles which make up my notes, the only legacy a meaningless litany of picturesque names – browntop, creeping red fescue, Cumberland turf.

Last week, sitting in front of the pavilion looking across to the Radcliffe Road end on one of the few days this summer when there was no play, all the terms came back to me: thatch, tillering, chlordane wormkiller, Sisis Rotorake. I knew the vocabulary, at least, even if I could not speak the language.

The ordinary lawn will have five or six different grasses in it, some of them, like Yorkshire fog, distinctly coarse and undesirable. The square at Trent Bridge has two, basically: 80 per cent Chewings fescue (named after a New Zealand exporter of grass seed), 20 per cent browntop, also known as common bent; in addition, at both ends, between stumps and crease where wear is greatest, a pinch of one of the new perennial rye grasses which will withstand close mowing. It is thus entirely composed of very fine-bladed, strong-rooting grasses which can survive being cut to three-sixteenths of an inch and the tremendous compaction necessary in the making of a wicket. The groundsman is far less interested in the cultivation of blades of grass than in that of their roots, which will bind the soil together.

The process of consolidating the wicket by rolling begins in the very early spring. Mr Allsopp uses successively heavier rollers, finishing with a two-and-a-half-ton petrol-driven roller. (Gone are the days of rollers drawn by horses, noble animals like Horace who, in the 1890s at Trent Bridge, used to move towards the roller as the invariable tenth man, Fred Morley, left the pavilion for the crease.) This even compaction, to a depth of four to five inches, is the secret of a good, true wicket, and it demonstrates clearly the tenacity and resilience of the fine grasses which can tolerate such treatment. Ten days before the Test, Mr Allsopp measures out his twenty-two yards, mows closely, and floods the pitch with water. The next day, the heavy roller is applied for several hours until the surface is dry. The covers are then put on and there will be no more water on the wicket until the end of the Test, although it will be mown each day that the match is in progress.

The groundsman's season properly begins in the autumn, the moment the last ball is bowled for the summer and the last stumps drawn. It is in these autumn operations that it is possible to discern a similarity between groundsmanship and lawn maintenance. All the harm perpetrated by a summer's ceaseless pounding by boot, bat and ball and the furious rolling to keep the pitch hard and level is to be undone. To give the grasses a chance to recover the compaction must be relieved and the thatch (dead grass which accumulates in the season and undermines the even bounce) removed. A mechanical scarifier with sharp knife-blades cuts through the thatch, which is then swept up. Holes are spiked at regular intervals to let air circulate around the roots. The turf is fed with a low-nitrogen, slow-release fertiliser which encourages root growth in winter. Unlike most gardeners, groundsmen cannot co-exist peacefully with earthworms, so they are killed. Grass seed is sown where the wicket has become worn. After more laborious spiking with a border fork comes a topdressing of heavy loam, much of which is brushed into the spike-holes. (This is no longer the famous Nottingham marl, the clay once used in grounds all over the world, which, when rolled too much, transformed the wicket into a level Sahara.) The loam has to be heavy, for a light soil would easily break up under the punishment it receives in the summer. The weeds of our own lawns – daisies, dandelions, plantains – are killed with herbicides or mown out.

The unhurried perfectionism of the groundsman makes most people's half-hearted efforts at lawn-care look very puny. The autumn maintenance of a cricket square, however, tackles much the same problems as beset lawns. The wear that most children will exert on grass is comparable in kind if not degree. We gardeners, having exchanged the hard work but beautiful effect of the cylinder mower for the easy life of the rotary or hover mower, now go to great lengths to avoid the scarifying, moss-killing, rolling, spiking and top-dressing which the use of those mowers makes even more necessary. The quality of most lawns has rapidly deteriorated. I used to laugh at lengthy and exhaustive descriptions of the jobs required to make a decent lawn, but as I stood and looked at the wicket to be used for the Test, with its neat, dense, uniform fine grasses and oh-so-level surface I finally saw the point. We are terrified of the idea of being slaves to our grass. It is refreshing to find at Trent Bridge such a bondage cheerfully, even enthusiastically, embraced.

About ten years ago, a collective decision was taken at Nottingham to alter the pitch from a slow, batsman's wicket to one more likely to produce results. This was achieved partly by withholding water from the pitch for a longer period before the match, so that the surface was drier, the fast bowlers would find the pace they needed and, when the shine went off the new ball, the spinners would find the ball gripped and turned on the practically bladeless pitch. This policy required courage, for there is the ever-present fear that the playing surface will begin to break up before the five days of a Test Match are over, although up till now this has not happened. Mr Allsopp has also experimented with a lighter top-dressing in the autumn, but he freely admits that the making of a wicket is not an exact science, and that two wickets on the same square, prepared in the same way, may play differently. One thing undoubtedly makes a difference, and that is the TCCB* ruling that first-class wickets must be covered, when play is over, for the duration of the match. The surface inevitably dries as the game progresses and, if the sun shines, the 'life' may well go out of the pitch as the soil below also loses moisture. In the days when wickets were not covered, rain could soak in and, if the sun came out, the result was the famous sticky wicket which so confounded batsmen. The conditions then could be blamed on the weather, but these days the first-class groundsman must cultivate a hide as hard as his square for the occasions when the visiting side lose.

Mr Allsopp seems cheerfully resigned to the criticism which is his lot as a head groundsman. He knows how many elements which are completely beyond his benign control will influence the outcome of the Test this week. He also knows that

*Test and County Cricket Board, as it was called then.

a batsman's 'gardening' is far more likely to be the result of nerves than the presence of bumps or divots on his meticulously gardened chain of turf.

Green Touch   *20 March 1999*

We are rugby football fans in our family, so fanatical that we could make a whirl of Dervishes look half-hearted. We support the England team, of course, but we care even more about our local one, the Northampton Saints. Apart from the rugby, we enjoy mixing with such conspicuously loyal, knowledgeable and well-behaved supporters: so courteous that when the opposition kicks for goal, you can hear a match struck on the other side of the field; so gloriously partisan that we take it as a personal affront whenever a Saint – Rodber, Dawson, Grayson, or Beal – is dropped from the England team.

Waiting for the players to run out on the pitch at Franklin's Gardens, I have often reflected on the quality of the turf on which the game is played. It is an admirably green, thick mat of healthy grasses, mown surprisingly short. The man with responsibility for this turf is the Grounds Manager, David Powell, once a famous prop forward who won 23 caps for England, toured with the Lions in 1966, and made 370 appearances in the green, black and gold jersey between 1963 and 1977. He is just as much a hero to me for the quality of the pitch he produces, however.

It is not easy for him. Rugby players put more strain on grass than any other sportsmen. It is not hard to imagine what sixteen stud-booted, scrummaging bodies, in total weighing three and a half tons, can do to turf. And, in this enormously demanding professional era, they are doing it twice a week for more than eight months of the year. Yet the pitch almost always looks in fine order.

As you would expect, David Powell is as solid as a block of

Northamptonshire ironstone, with a ruddy face and ears individually shaped by years of packing down in the front row. As
befits a man who has earned his rugby stripes, he is quietly
confident and sure of his art. He is also a farmer, and his shed
of machinery reflects a farmer's practicality. His irrigation
pump, for example, is an ingenious graft of combine harvester
engine and modified 'dirty water' pump.

It is said (in Northampton at least) that he prepares the best
rugby pitch in England, which explains why he is presently
not a happy man. The entire annual average amount of rain,
amounting to 36 inches, fell between November and the end of
January this year, and March is turning out to be very wet also.
He has been forced to admit that, for the first time in eight
years, the pitch is sufficiently damaged to need re-seeding. He
has three short summer months, from May to August, in which
to achieve a new playing surface. It will mean taking off the top
three inches of grass and sandy soil, mole-ploughing to a depth
of four inches, and then incorporating a sand and fibre mix into
the top inch. The theory is that this will stabilise the playing
surface while ensuring that it is not so hard that players will
injure themselves, nor so soft that they cannot run through
it. Once done, 'pre-germinated' grass seed will be sown at a
rate of an ounce and a half per square yard, a rate we gardeners would consider very extravagant. The mix is composed
of 'perennial ryegrass', 'smooth-stalked meadow grass' and
'slender red fescue', together with 'tufted hair grass', which
grows well in sand and will also stand some daily shade from
the covered stands.

In essence, however, the groundsman's task is no different
from that of the stock farmer or gardener. It is a matter of
growing healthy grass by using suitable seed and providing
good drainage and plenty of air round the roots. David Powell
may choose a very hard-wearing seed mixture, and must perforce sow later in the year than any gardener would contemplate, but he is still making a gigantic lawn.

His groundsmanship is surprisingly 'green'. The water which drains off the pitch in winter is collected in a lake nearby, then extracted and put back on the field in summer. He cannot remember when he last used any weedkillers, and he stays his hand with most fertilisers, too, especially nitrogen. In fact, it horrifies him to think how much we gardeners routinely put on our lawns, even when we have children and pets. Times have changed, in this as in so many aspects of rugby: in his playing days, cuts almost always went septic.

Maintenance is mechanical, not chemical. The way that his two assistants, Paul Johnstone and Steve Robinson, keep the turf free of weeds and thatch, and the soil aerated, is to cut through the grass roots with a slitter machine twice a week. They also mow every day, letting the clippings fly, but sweep with a brush once a week. We may blanch at that kind of commitment, but David believes that garden lawns could be so much better if we did the equivalent of slitting; that is, if we spiked them often with a four-tined garden fork to a depth of four to six inches. Our lawns look sickly yellow in the spring, not from want of nitrogenous fertiliser so much as from lack of air in the topsoil and good water percolation through the root sphere.

On Saturday, when I watch England (including, I trust, a sprinkling of Saints) playing France at Twickenham on television, I shall think of David Powell and the other club groundsmen who somehow must produce a suitable and sympathetic playing surface more than forty times a year. As for my own lawn, I intend to take good heed of his advice – once the final whistle has blown.

Present Time

11 December 1999

If gardening had not existed, we should certainly have had to invent it. How else could we grapple with those eternal questions like what to do with a Sunday afternoon in November or, even more important, what to choose to give our relations and friends for Christmas? Gardening, or its associated hardware at least, can prove our salvation in early December, giving us the leisure to concentrate on the truly everlasting aspects of the festival, safe in the knowledge that we have carried out the least important, and most time-consuming, of all Christmas duties.

All the world loves a gadget. As every parent knows, the most enjoyable presents to buy for children are those which promise fun or intrigue. We never grow out of that. The garden centres and mail-order catalogues are stuffed with knick-knacks – terracotta labels, ornamental pot 'feet' and cane tops, radar frogs which croak at you alarmingly as you walk by, house plant waterers in the shape of hollowed-out kingfishers – each devised to make the most mundane horticultural milieu slightly less dull. There are any number of handy-looking little tools for digging weeds from paving, sharpening knives, wrapping outside taps, cleaning boots, sowing seeds, all holding out the possibility of a slightly easier, or certainly more agreeable, gardening life. The ubiquity of these 'gift ideas' is proof of their popularity.

In the end, however, all this is froth and bubble. If you really want to give your sister-in-law something for which she will be truly grateful and remember you with kindness every time she uses it, then I am afraid that you are going to have to spend some serious money. The most useful and acceptable gardening presents are undoubtedly those which will set you back a bit: a pair of swivel-handled Felco secateurs, for example; a garden vacuum cleaner (yes, I mean it); a trolley truck; a family of stainless steel tools; a potting bench; a 'quiet' shredder; a weather station. If you are after something more aesthetic, there are any number of pigs made out of willow, hammocks, 'verdigris' cranes, armillary spheres or Versailles tubs on the market. If you persist in giving her a T-shirt with 'Gone to Seed' stamped on it, and she reads this, she will know just how much thought you have given to the matter.

153

For some reason, none of my enormous family, except for my garden-minded father and my ungarden-minded but thought-ful father-in-law, ever gives me anything vaguely horticultural for Christmas, unless I specifically ask for it, and even then only after a lot of fuss and cries of 'how boring'. I suppose they think that I have everything, which is very far from being the truth. In any event, just because I have a pocket knife, it does not mean that I could not use another one. I have, after all, more than one pocket.They must think I am like the man who was asked if he would like a book for Christmas, and replied 'No, thank you, I already have one.'

As with books, so with plants. I have more than an acre of garden, not every inch of which is crammed to bursting with something delicious. But even if it were, I am a gardener and, *ipso facto*, always prepared to fit one more plant in. How I should like someone to give me a plant that they admired and grew, and had bothered to consider whether I might like it as well. Wouldn't you?

As a nation, we probably don't go in enough for giving plants at Christmas (except for bright, jolly indoor cyclamen and blood-red, baleful poinsettias, of course), either because it doesn't seem the right time to be planting outside, although it is perfect for anything bare-rooted, or because we think that even gardeners have other things on their minds in late December than digging holes. That may be true, but the beauty of ordering plants now is that they will arrive some time after the New Year, when Christmas is a distant memory and we are all in need of cheering up.

If this idea appeals to you, I suggest that you concentrate your thoughts on relatively expensive but straightforward plants, such as most hellebores, unusual snowdrops, new roses, or camellias – the sort of plants the receiver might not feel they could justify buying themselves.Then ring, write or e-mail for the catalogue of an appropriate mail-order nursery listed in this year's edition of the *RHS Plant Finder*, choose your plants, send off an order, and alert the recipient at Christmas with a card. One day in late winter or early spring, when he or she has quite forgotten all about it, an interesting parcel will arrive, full of latent but potent promise for the garden, not just for this year but for many years to come. You could hardly claim that for bath salts, could you?

7

Gardening Types, Personalities and Organisations

Spinners of Dreams   *21 February 1998*

If you want to know why there has been such an explosion of interest in gardening over the last decade or so, look no further than the refinement of garden photography, together with the improvement in the quality of colour printing. The enormous number of high-class, highly-illustrated gardening books and magazines now available owes a great deal to the skills of the modern garden photographer. Never have images been sharper, more beguiling or, in PR-speak, more inspirational.

Indeed, so good are the pictures and the colour origination these days that gardening features are now prominent in colour supplements and general-interest magazines as well.

Not surprisingly, many people who might well have resisted the allure of what used to be known as 'chatty' gardening books (those with plenty of text and perhaps a few line drawings) have found illustrated books much more to their taste. Even those of us who were already converts to gardening books have marvelled at how much superior pictures have added to our understanding and pleasure.

Demand has created a ready supply of professional garden photographers, even though the job requires very particular qualities. Apart from having the obvious, though unusual, combination of technical acumen and artistic sensibility, they must be tireless and meticulous, with a patience which Job might have found excessive. In the summer months they move ceaselessly from garden to garden, here and abroad, sometimes with a caravan hitched to the back of the car so that they can sleep on the spot and be up before dawn to catch the best light. Long before the garden owner is astir, they will be making a ghostly progress around the garden, noiselessly slipping between shrubs, leaving no footprints in the border soil, the only signs that they are human rather than ethereal being the step-ladder, tripod and box of lenses which accompany them everywhere.

Even in winter they are out very early, catching the rime on seed heads before the rising sun can melt it, or snow on the horizontal branches of *Cornus controversa*. They seem able to transcend the discomfort of being motionless for long periods outside in all weathers and at all seasons. Yet, despite this almost obsessive care and concentration, garden photographers are remarkably good-natured and clubbable. They sometimes need to be: persuading a frosty old-school dowager to allow them the freedom of her garden requires tact and social ease.

In order to make the most alluring picture, a photographer will naturally remove any blemish from a composition. Slug-chewed flowers, random fallen leaves, an ugly building beyond the garden: these are studiously removed, or hidden by artful camerawork. It is a matter of professional pride to do so and anyway, if they did not, their pictures would not sell. The effect of light is changed or intensified by the use of filters, lens hoods, reflectors and flash-guns. Even where the subject is neglect and decay – a clump of nettles by a broken-down wall, say – the picture will somehow be invested with a potent romance. Flowers shot in the rain always seem to glisten, rather than hang sad and sodden. No garden, no flower even, looks quite so perfect as a skilful photographer's pictures can make it.

Because of this, photographers – and the art editors who commission their work – are sometimes criticised for providing – and expecting – images which are 'unreal'. In one sense, this is self-evident. A frozen two-dimensional image of a moving three-dimensional scene cannot possibly be 'real'. But what is meant is that the image may create a deceptive impression. Photographers would argue that it is often our own perceptions which deceive, our imaginations which exaggerate or refine. Moreover, the camera can reveal detail, even the essential nature, of a plant or garden which the more careless eye of the spectator does not divine. And the frame of a picture inevitably gives a coherence and compositional unity to an otherwise more formless scene.

I will buy most of that: but in the case of blown-up pictures where plants are portrayed more than life-size, or when a wide-angle lens distorts the dimensions of a garden, I suspect I am being taken for a pleasant ride. I suppose I should mind more than I do. Photographers may be spinners of dreams, but gardeners are willing dreamers.

Novel Designs   *14/21 December 1996*

When Mr Darcy dived into the lake at Pemberley in the BBC seri-
alisation of *Pride and Prejudice,* the nation (or that half of it which
is female, at least) gasped. But was it the sight of Colin Firth's
chest, or was it disbelief that there could be an ornamental lake at
Pemberley? After all, as every Janeite knows, Pemberley was not
a Brownian 'landscape'. Mr Darcy owned a trout stream, which
he offered to put at the disposal of Mr Gardiner, but no lake.

This may not have mattered to every last viewer, but it would
surely have mattered to Jane Austen, whose novels are shot
through with references to contemporary attitudes on gardens
and their making, on landscape, nature and much else. Indeed,
each novel characterises, and often satirises, some development
in thought or fashion in Georgian England: the Gothick imagination in *Northanger Abbey*; the way landscape and nature were
viewed by devotees of the 'Picturesque' movement in *Pride and
Prejudice*; the role of the responsible landlord (Mr Knightley in
Emma); the development of the Reptonian Regency garden (Mr
Rushworth's desire for 'improvements' to the grounds at
Sotherton in *Mansfield Park*) and the rise of 'romantic' ideas, in
Persuasion. I know this because I have just read *Jane Austen and
the English Landscape* (Barn Elms, £19.99) by Mavis Batey,
President of the Garden History Society and expert on land-
scape gardens and their place in literature.

Anyone who loves Jane Austen's novels will probably find
that this erudite but accessible and attractive book adds to their
fun. To know, for example, that William Gilpin, pioneer of the
'Picturesque', suggested that three cows and no more were the
ideal number in a landscape composition is to understand the
joke implicit in the Netherfield shrubbery scene in *Pride and
Prejudice*. As Mavis Batey puts it:

When Miss Bingley and Mrs Hurst, who have been less
than civil to Elizabeth Bennet, are walking with Mr Darcy

... and he invites her to join them, she says pointedly, 'No, no, stay where you are, you are charmingly grouped and appear to uncommon advantage. The picturesque would be spoilt by admitting the fourth.'

Mrs Batey also neatly connects contemporary preoccupations, described or hinted at in the novels, with what is known of the lives and attitudes of the cultivated Austen family.

It is coincidental and fortuitous that this book should have appeared at the height of the present craze for Jane Austen, but Mrs Batey has been working on it for many years. She is a natural researcher, having begun a dissertation at university on the German Romantics (the same subject, she says, as Dr Goebbels'); this she abandoned after war broke out, to become what she calls a 'trained ferret', helping to decipher Enigma messages at Bletchley Park.

Her interest in the history of gardens grew from work for the Council for the Protection of Rural England in the 1960s, as well as from discoveries made when living in the agent's house at Nuneham, near Oxford. After a great deal of ferreting she concluded that Nuneham Courtenay, moved to make way for Lord Harcourt's landscape garden, was the model for Oliver Goldsmith's *The Deserted Village*. From 1971 to 1985 she was Secretary of the Garden History Society (founded in 1965 as a learned society), and has been involved from the beginning in its attempts to conserve and protect historic parks, gardens and landscapes.

One of the society's achievements has been its collaboration with English Heritage on the Register of Parks and Gardens of Special Interest. Even better, since 1994 planning authorities have been required, by law, to consult the society when they receive a planning application which concerns a park or garden on the Register. These days, the GHS is also often consulted by those wishing to make applications for Lottery money. Most important of all, in thirty short years and from a standing start, the conservation of historic gardens has become accepted as desirable and achievable, by public authorities and garden owners alike.

Mrs Batey has been an indefatigable lecturer, organiser of conferences, campaigner and writer of numerous academic papers and several popular books. Her personal contribution to the acceptance of garden history research as a respectable academic discipline, which now also has substantial practical implications for historic gardens, was recognised this year with the publication of a *Festschrift** to celebrate her seventy-fifth birthday. In one of the essays she is quoted as saying: 'I hope that as many people as possible will visit and love gardens, and that their history will become as much part of our lives as poetry and painting.' That day may not be too far off, the way the Garden History Society and its energetic President are going.

Going Native   *21 March 1998*

Like most other people, I marched in London on 1 March 1998 to lend my support to field sports. Why? Because they play an important role in the conservation and well-being of wildlife and the countryside, not to mention country people. I could do no less, since gardening, likewise, has played its part.

The steep decline in numbers of many birds, small mammals, amphibians, butterflies, moths and insects, reported in scientific papers but reinforced by the evidence of our own eyes, fills us with real dismay. In the case of most birds, butterflies and insects this is directly related to the scarcity of the wild flora which sustains them; for they have, in the jargon, 'co-evolved', and many have specific tastes. The

---

*'Essays in Honour of Mavis Batey' in *Garden History*, the Journal of the Garden History Society, Vol. 24, No. 1 (Summer 1996).

brimstone butterfly, for example, lays its eggs only on buck-thorn, *Rhamnus catharticus*, a shrub of hedgerow, fen and woodland.

So I consider myself fortunate that, with no wages to pay or markets to satisfy, I have nothing to lose by providing shelter, safe breeding sites and suitable food plants for wildlife in my garden, and everything to gain in enriching experience, not to mention a delicious smugness. As if that were not enough, I can do it without compromising unduly the conventional notion of what an attractive garden should be.

In fact, anyone can, without giving up their attachment to precious exotic plants. Those with large gardens should be able to find space for an extensive patch of nettles where red admirals can breed, long grass for insects and small mammals, a spot of poor soil for wild flowers, a pond, a mixed hedge of native species, and even perhaps a few deciduous trees (a mature oak can support 240 species). Even in a tiny garden, something may be achieved by growing single rather than double nectar-rich flowers, planting wild flowers and a berry-ing shrub, and allowing both cultivated plants and wild ones (in selected places) to set seed. And, unlike field sports, all this can be pursued in towns and cities, too – and, what is more, be met with almost universal approval.

The premise that native plants are the key to attracting and keeping a diverse fauna underpins the work of a recently formed charity called Flora-for-Fauna. Based at the Linnean Society in Burlington House, Piccadilly, it aims to encourage people to grow a range of locally indigenous plants which are part of the complex web of dependency connecting insects, butterflies and moths, small mammals and birds.

Flora-for-Fauna is the brainchild of Jill, Duchess of Hamilton, an Australian writer long settled in Britain whom I first met when she was putting together a show-garden of Australian plants at the Chelsea Flower Show in 1994 (her name is a legacy from a past marriage to Scotland's premier peer). She is smart in every sense, combining energy and brains with a glamour which is far from commonplace in the botanical world. These attributes have earned her some influential friends, among them John Simmons, past curator of the Royal Botanic Gardens, Kew; Professor John Parker, director of Cambridge Botanic Gardens; and Miriam Rothschild,[*] who has pioneered the creation of wild flower 'meadows' in gardens, on road verges and in parkland, using nursery-grown seed.

Jill Hamilton has succeeded in persuading a number of businesses (Rio-Tinto, Kleinwort Benson, Osborne and Little, Royal Mail) to part with money. There are several educational projects presently under way, including research in Cambridge on nectar plants, but the one which has attracted most attention is the Postcode Plants Database; it was ingeniously, and laboriously, assembled from local flora and fauna distribution maps with funding from Royal Mail.[†] Anyone online may tap in their postcode and a list of the local wild flowers, butterflies and birds will appear on the screen. I was suitably humbled by the number of excellent garden plants which I don't grow.

One virtue of cultivating locally-indigenous plants, according to Flora-for-Fauna, is that they have evolved to take advantage of regional soils and climate, so they should require little maintenance. But then, neither does buddleja, a Chinese plant much patronised by butterflies. However, as Jill Hamilton points out, buddleja, though important for its nectar, is not a food plant for caterpillars. Moreover, she is not suggesting we grow just native plants, only that they should form an important element, perhaps a third of the plants in every garden.

Possibly to counter the charge that wild flower gardens can be scruffy, a 'formal garden' is planned for Chelsea Flower

---

[*]Now Dame Miriam Rothschild.
[†]The Postcode Plants Database may be found at fff.nhm.ac.uk/science/projects/fff/

Show this year. Its centre-piece is to be an elegant twenty-four-foot high octagonal, three-tiered tower, made mostly of wooden trellis, with special niches for bats and different birds. 'The Birds' Buffet' (how Australian that sounds) tower will be surrounded by indigenous plants and hedging, and a pond. It has been created by the ultra-fashionable garden designer George Carter, and will be sponsored by Christie's. Going native looks like being the smart thing to do this year.

Hereditary Principles   15 May 1999

I am an enthusiastic supporter of the hereditary principle.* This naturally disposes me in favour of the peerage. But it also puts me on the side of the breed of traditional professional gardeners who pass their skills, their interests and their attachment to continuity to their descendants. The three generations of the Puddle family who have been successive Head Gardeners at Bodnant in North Wales are only a particularly striking example of the virtues of heredity.

My prejudices received a resounding endorsement at the latest Westminster Show of the Royal Horticultural Society. The Duke of Devonshire marked publicly, with a large exhibition of glasshouse plants from Chatsworth, the retirement after fifty years' service of his Head Gardener, Jim Link, and the appointment of his successor, Ian Webster. Jim Link is the son of Bert Link, who also clocked up fifty years in the gardens there, finishing as Head Gardener. Between them came Dennis Hopkins, who also stayed for fifty years and whose retirement was similarly marked by his employer in 1989.

The plants at Vincent Square were from the Duke's private greenhouses, which exist to provide floral decoration for the

*This article was written at the time when the Government was concerned with abolishing the hereditary element of the House of Lords.

house: there were orchids, gardenias, fragrant tender rhododendrons, *Lilium longiflorum*, amaryllis, camellias, and a centre-
piece of *Musa cavendishii*, the banana which bears the family
name and produces fruit in the glasshouse at Chatsworth.

Jim Link worked initially at Chatsworth as a forester,
moving to the gardens in 1981 and becoming Head Gardener
in 1989. He also knows a great deal about drains and pipework,
important for the smooth operation of the cascades and fountains. Under his tutelage there have been a number of substantial improvements, including the renovation of the kitchen
garden, although he considers his best achievement to have
been the removal of the terribly invasive *Rhododendron ponticum* and laurels and the programme of replanting in the abore-
tum, pinetum and Stand Wood. When I asked about his time
at Chatsworth he said, without affectation, that working there
had been a privilege. Like Ian Webster, he has the confident,
cheerful demeanour of one who knows he can do his job, and
is valued for it.

The latter, also a Derbyshire man, has been employed at
Chatsworth for twenty-seven years, and looks after the glass-
houses. It is he who competes with the Head Gardener from
Blenheim for first prize in the glasshouse grapes class at the
RHS show every October. The honours are slightly in the
Devonshire favour, although the Duke of Marlborough's
grapes won last year, a fact which Ian, not unnaturally, puts
down to the declining vigour of Chatsworth's eighty-year-old
'Muscat' vine.

Although it is not a view to please the *bien-pensants*, it seems
to me that, taken all in all, it is most satisfying to work as a professional gardener for a single boss – and the grander and more
long-established, the better. I still hanker, many years later,
after my time in 'private service'; contrary to popular myth, I
was never patronised or exploited, no one expected me to be
deferential, simply polite, and the work was endlessly fascinating. In places where employers take a personal and careful

interest, as they plainly do at Chatsworth, an easy, mutually respectful, and creative working relationship is not only possible but likely. This must, at least partly, explain the loyalty of the Chatsworth men.

The problem, of course, is finding gardens big enough to warrant employing more than one gardener, or owners able to afford it. Gardeners are usually social animals; like horses, they rarely thrive on solitude, though many are forced to try. There are twenty-one gardeners at Chatsworth, with six more in the park and woods, which makes for a far bigger, and probably jollier, workforce than you will find in most landscaping firms.

Besides company, what matters to professional gardeners almost as much as to garden owners is continuity. The crunch often comes when a house is sold, but this is hardly an anxiety for anyone at Chatsworth, which is in demonstrably better case now than when Jim Link started work there soon after the last war. The best gardens develop and change over the years, and the happiest gardeners are those who know that some at least of their work will endure. If the hereditary principle operates so well in the exacting world of professional gardening and garden ownership, it is rather hard to see why it should not do so in the comparatively simple sphere of Parliamentary government.

Il faut cultiver notre jardin   *17 October 1992*

I have wished, when burying my daffodil bulbs in the ground this month, that I might bury my head along with them. An understandable reaction in these troubled times – and one shared by several of my friends, especially the women. Mention the future of Britain and Europe* and our ears go back like the petals of *Narcissus cyclamineus*. In this unreconstructed

*This was the time of the Maastricht Treaty negotiations.

household, worrying about the legal validity of 'subsidiarity' is definitely man's work.

Unable to escape the avalanche of news, I too have had my anxieties, even if I choose to dissipate them in enthusiastic leaf-raking rather than by growling my way through the Sunday newspapers. Conscious that it is easy to get things thoroughly out of perspective, I have taken to reading, as a corrective, old issues of the *Journal of the Royal Horticultural Society* from the Second World War. Having been born years after hostilities ended I do not know what really anxious times are, so this has proved salutary.

The *Journal* has always been both symbol and expression of the horticultural world at large, but hardly of the world beyond. What immediately struck me was how well people appeared to be able to forget international troubles in writing about camellias flowering in Cornwall or the results of the delphinium trials at Wisley. But if it was escapism, it was escapism of the most positive kind.

Initially, the war seems to have come as an unexpected shock to the RHS. There is no mention of European difficulties in the September 1939 issue (Germany invaded Poland on the lst, and war was declared on the 3rd). The Great Autumn Show, due to be held in the middle of that month, was only cancelled at very short notice. However, the tone of the October *Journal* (the British Expeditionary Force was landing in France) was already upbeat. The Secretary could write: 'The Council is very anxious that the necessary increase of vegetable crops shall not entirely displace the ornamental aspects of our gardens. The mental refreshment and recreation which flowers can give us all will be a notable contribution to the national spirit . . .'

By May 1942 (the Japanese took Mandalay on the 2nd), the cardboard covers of the *Journal* had been abandoned and it had shrunk in size and number of pages. Yet, despite paper shortages, the *Journal* continued to be published throughout the war, sent out monthly to thousands of members, then called Fellows, and sold to non-Fellows for 1*s*. 6*d*.

The war does intrude, to be sure. There are, for example, touching letters from British officers in prisoner-of-war camps, reporting how pleased the men were with the flower and vegetable seeds sent by the Society. One complained, however, that as the camp had been pine forest until recently, the soil was poor and sandy and did not provide ideal growing conditions. The Society also organised a sale of orchids and other plants for the Red Cross.

Underlying the elegant little essays about the founders of the Society and meaty accounts of recent research into lily propagation is a desire to ensure that, whatever happened, horticultural life and, in particular, scientific investigation should not be allowed to fizzle out. Lectures and meetings continued to be held. Even flower shows survived, albeit on a smaller scale: only Chelsea Show was a casualty of the wartime uncertainty, not being held again after 1939 until 1947.

To those who remembered the pre-war *Journal*, the wartime effort must have seemed a shabby, makeshift affair, but from this distance the wonder is not that it was done badly, but that it was done at all. I imagine that its continued existence owed much to the Ministry of Agriculture's need to encourage amateurs to grow their own food, but that hardly explains the presence of a scholarly article on graft hybrids and chimeras the month after the evacuation from Dunkirk in June 1940. There was, interestingly, also a note about the increase of Colorado beetle in western France, where there were 'abnormal conditions at present prevailing . . .' Even the Blitz in May 1941 (the Chamber of the House of Commons was reduced to rubble on the 11th) did not prevent a small flower show from being held at nearby Vincent Square in June.

Such bravado springs from a most understandable desire not to feel entirely without influence over the world around us, or let large, uncontrollable events spoil all our fun. For me, these wartime journals forcefully if quietly make the point that, whatever happens, *il faut cultiver notre jardin*.

Scot with Courage   17 July 1999

I have a particularly soft spot for David Douglas, the nineteenth-century plant hunter: because he was the best kind of clever, energetic and adventurous Scot, in the John Buchan mould; because we share a birthday (his bi-centennial is celebrated this year); and because he found and described one of the all-time great shrubs, *Dendromecon rigida*.

Of course, it is not a rare, slightly tender, glaucous-leaved, yellow-flowered, fragrant shrub, however lovely, on which Douglas's reputation rests. Rather, it is on the introduction of a number of invaluable garden plants – the flowering currant, *Mahonia aquifolium*, lupin, penstemon, oenothera, erigeron, clarkia, and eschscholzia – together with a clutch of hardy coniferous trees – the Douglas fir, the noble fir, the grand fir, the Sitka spruce – whose commercial and ornamental attributes have ensured that they are widely planted in Britain. It is no exaggeration to say that Douglas's collections have succeeded in radically changing the landscape of upland Britain. Not many people will be happy from an aesthetic point of view that those trees are conifers, but their value as timber is indisputable. Moreover, in Australia and New Zealand the vast majority of planted timber is composed of the Monterey pine, another Douglas introduction.

David Douglas was born in the village of Scone in Perthshire on 25 June 1799. He showed an early interest in natural history, and at the age of eleven was employed as a garden boy by the Earl of Mansfield at Scone Palace. By 1820 he had studied so hard, and acquired so much experience, that he was taken on at the Botanic Gardens in Glasgow, where he soon impressed the Professor of Botany, William Hooker, one of the great botanists of his day. Hooker recommended Douglas to the Horticultural Society of London (now the Royal Horticultural Society) as a plant collector. And so it was that in 1823 he travelled to the eastern states of North America, and the following year, via Cape Horn, to the Pacific North West. He stayed there for two and a half years, living among native North Americans and travelling vast distances on foot or in canoe. In 1829 he set sail again from England for Hawaii, and from there to California, on an expedition which was to last five years. On more than one occasion his canoe was overturned and he lost seeds, plant specimens and instruments.

Eventually he returned to Hawaii, but was killed in July 1834 when, on the slopes of the volcano Mauna Kea, he fell into a trap for wild cattle which already contained a bull. He was buried in Honolulu. His tenacity, dedication, hardiness and courage in wild places, coupled with an extraordinary eye for plants which would prove useful in Britain, made him one of the most successful collectors ever. In all, he introduced 249 new species, of which 130 proved hardy here.

Unlike some of his contemporaries, such as Archibald Menzies, who actually first found the Douglas fir (which is why it is called *Pseudotsuga menziesii*), he is still widely remembered. In the summer of 1999 there were a number of celebrations, at Scone Palace and elsewhere. Many of the events were organised by the David Douglas Society, founded in 1997.

Despite his obvious distinction, I have a feeling that the nature of his going is an important reason why Douglas is still well-known. After so many hardships, privations and adventures in his short life, to fall into an animal trap seems both desperately sad and shiveringly horrific. His tombstone in Honolulu bears an epitaph taken from Virgil: 'Tears are due to wretchedness, and mortal woes touch the heart.' Indeed they do but, as Virgil has Aeneas say of Dido, 'While the rivers shall run to ocean, while the shadows shall move in the mountain valleys, while the sky shall feed the stars, always shall thy honour, and thy name, and thy glory abide.'

Rex Crocorum *21 February 1987*

Crocuses, like sorrows, come not single spies, but in battalions. One February day there are none, the next morning dozens have nosed through the short grass. Winter is not yet vanquished, but a strategic withdrawal is in progress. All crocuses are worth growing if, and it is a big 'if', they are left alone by the birds, but none more so than the varieties of *Crocus chrysanthus*.

Smaller than the conventional garden crocuses, at barely three inches tall, these are the prettier for the varied markings each bears upon the outside of its petals. Almost without exception, crocuses open out only in sunny weather, so these markings are most valuable in the usual run of dull winters. 'Ladykiller' is purple-blue, and white inside; 'Blue Pearl' blue within, almost white without; while the unusual 'Zwanenburg Bronze' is feathered and stippled with bronze overlying yellow. Best of all, for its large butter-yellow flowers which are bronze at the base, is the variety named in honour of 'the Crocus King', E.A. Bowles.

For some reason, connected perhaps with his reluctance to show off, Edward Augustus Bowles is not well-known beyond horticultural circles. A saintly gentleman of leisure whose family had made their money from the New River Company which brought clean water to London in the seventeenth century, he spent his long life giving service to the Royal Horticultural Society and the poor of the district of Enfield in Middlesex, and gardening with a botanist's clear eye and a gardener's feeling at his parents' home, Myddelton House, Bull's Cross, near Enfield. He had no conspicuous love affairs, won no prizes for poetry, and fought in no wars. Yet [in 1987,] more than thirty years after his death, he is still remembered by gardeners with great affection.

His reputation depends partly on the myriad good garden forms of plants which emanated from Myddelton House; not only crocuses like 'Snowbunting' and 'Yellowhammer' but 'Bowles's Black Pansy', *Penstemon* 'Myddelton Gem', *Iris reticulata* 'Cantab', and *Milium effusum* 'Aureum', 'Bowles's Golden Grass'. He is remembered also for the trilogy of books published during what he called 'The Kaiser's War': *My Garden in Spring, My Garden in Summer, My Garden in Autumn and Winter*. The three books are shot through with a cosy, scholarly humour which makes them readable more than seventy years after they were written. Bowles's trilogy is as enjoyable as any work of twentieth-century horticultural literature and, considering the botanical detail included, one of the most accessible to the layman. He wrote to entertain as well as inform:

> One little yellow Crocus has an obnoxious trait in its character, and is a little stinking beast, as Dr Johnson defined the stoat. It is well named *graveolens*, and its heavy scent is generally the first intimation I get of its having opened its flowers. Sometimes I get a whiff of it even before I reach the Crocus frame – an abominable mixture of the odour of blackbeetles and imitation sable or skunk, or one of those awful furs with which people in the next pew or in front of you at a matinée poison you.

In these books he describes his garden, where he cultivated, besides a large collection of rock plants and 135 kinds of crocus, a miscellany of strange plants in a bed called the 'Lunatic Asylum'. The 'corkscrew' willow, the double 'rose' plantain and the 'oak-leaved' laburnum found a place here, with a magnolia planted to act as 'keeper'. He was a 'curious' gardener in the tradition of Parkinson and Ray.

His career has a wider significance, however, than that

embodied in the rearing of good garden plants or the publication of a set of agreeable, discursive books and scholarly monographs. By example and influence Bowles, together with like-minded contemporaries such as Reginald Farrer, encouraged gardeners into the ways of the 'plantsman'. I do not believe that the 'plantsman' tradition has ever been stronger than it is today, when the small size of many gardens encourages the plant specialist at the expense of the landscapist, but unfortunately the word is in danger of becoming so overused as to be stripped of all meaning. It is used by people to describe themselves if they know, or care to know, the difference between, say, *Daphne odora* and *D. jezoensis* and, at the same time, are uncertain how best to design their garden. To be a plantsman, one must grow plants for their own sake, in the places where they will be happy, but the term should presuppose a vast and intimate knowledge of individual plants, such as Bowles possessed, and should not be so universally and loosely applied. What it does not mean is the freedom to abandon all ideas of garden design – only that these should be accommodated, if necessary, to the needs of plants. Myddelton House garden, with the New River flowing through it flanked by beds of bearded irises and tulips, had a coherence and charm which did not depend solely on the individual beauty of plant species. Bowles did not rebel against the notion of colour schemes; he merely ensured that the well-being of his plants was never sacrificed to it.

The garden at Myddelton House has experienced mixed fortunes since Bowles's death in 1954. The famous rock garden has gone, as have the crocus frames, but the rest is presently enjoying a revival in the hands of the Lea Valley Regional Park Authority.*

---

*The garden is now fully restored and open on weekdays all year, and on Sundays in the growing season.

In Spite of All Temptations 17 August 1991

Last spring, while walking on one of the fells by the side of Ullswater, I found a primrose with yellow and green variegated leaves. Although these looked as if they had been inexpertly sprayed with weedkiller, I nevertheless badly wanted that plant. I was experiencing that sensation well known to all plant hunters: the desire to possess something strange or rare. Having been brought up not to pick most wild flowers, let alone dig them up, this caused a mighty but ultimately unsuccessful tussle with my conscience. I searched round for some means of uprooting it. I had reckoned, however, without the stern, unbending probity of my husband. With many a backward glance, I left the primrose behind, to blush unseen by all but the occasional rambler and incurious sheep.

I wondered how the great plant collectors of the past would have behaved, before everyone worried so much about conservation. Would Reginald Farrer, for example, have hesitated over an unusual primrose, except to admire it in extravagant terms, before getting out his trowel?

Born in 1880 into a Yorkshire landowning family related to the Sitwells, Farrer developed two rock gardens and an alpine plant nursery in the garden of his home at Ingleborough Hall and spent much of his short life on long collecting trips: in the European Alps, in Tibet and Kansu, China, and in Burma. Funded mainly by his family and by moneyed garden owners who expected seed in return for their investment, he also supplied botanic gardens, such as the one at Edinburgh, with seed, plants and some dried specimens. He died a lonely, courageous death while on an ill-fated collecting trip to the frontier ranges of Upper Burma in 1920.

What set him apart from other gentleman 'amateurs' with a scientific bent and independent means was his assured literary talent. He wrote five absorbing, witty and sometimes thrilling books about his travels and collections, as well as several on rock

174

gardening. The most impressive of these was his monumental two-volume *The English Rock Garden*, published in 1919. It is still prized by alpine enthusiasts – as much, it must be said, for the humour and extravagance of the prose as for the information conveyed therein. If he overwrote at times, it was out of exuberance and a desire to entertain. His approach was, for an amateur botanist, quite rigorous, yet he held fast to the layman's urge to be understood; his descriptions of plants are accurate, yet often truly poetic. Best of all, he could communicate his extraordinary affinity with mountains and their flowers, and so transport his readers to the 'high and lonely places'. He also introduced some lovely garden plants, among them *Viburnum farreri*, *Buddleja alternifolia*, *B. crispa* var. *farreri* and *Meconopsis quintuplinerva* – as well as a great many which have proved impossible to keep in cultivation for long, such as *Saxifraga florulenta* and *Farreria pretiosa*.

These days we feel happier with the idea of plant-hunting expeditions which are led by professional botanists working out of an international network of scientific institutions, rather than those of nurserymen or private individuals collecting on their own account or only loosely for the benefit of botanic gardens. Nevertheless, botanists and gardeners alike owe Farrer an enormous debt: not only for his introductions and for his books, but also because his posthumous influence and reputation prompted the foundation of the Alpine Garden Society in 1929.

By the time this is published I shall be in the Alps, not far from Farrer's favourite stamping grounds and, I hope, doing some gentle botanising. It may come as a relief to you to know, however, that all I shall be taking with me is a field guide, a hand lens, a camera – and my husband.

Frank's Fame   22 *January 2000*

These have not been good times for the famous. Many VIPs must have rued the day they were flattered into leaving their fireside to pass the last evening of the old century in the Greenwich Dome, since in the end they spent most of it in a queue in a draughty Underground station. And to cap it all, they were later dismissed as 'toffee-nosed' and 'snobby' (the ultimate indignity for chaps like Greg Dyke, surely?) by a former Minister of the Crown.* On the first Sunday of the year the Archbishop of Canterbury took the Sermon on the Mount as his text, a passage in the Bible guaranteed to make any VIP squirm. Some days earlier George Harrison, a man who long ago recognised that fame amounted to dust and ashes and took himself off to enjoy a relatively private life, was stabbed at his home, Friar Park near Henley-on-Thames. Who said fame is as evanescent as a bubble? For George Harrison, it has proved all too enduring.

Interestingly, Harrison is not the first famous person to have lived at Friar Park. Garden historians will tell you that the builder of the house and garden was a rich Victorian swell whose name became widely known, in horticultural circles at least, as a result of the so-called Crispian Row. Sir Frank Crisp was a City solicitor of standing and influence, a generous man to his friends, and a strong adherent of Anglo-Catholicism. He commissioned the 120-room Friar Park in the Gothic Revival style in 1896. The enormous garden was divided into a number of nostalgic period elements: there were, *inter alia*, an Elizabethan-style garden, a Dutch garden, a herb and 'nosegaie' garden with plants chosen from sixteenth-century literary sources, a 'Boccacio' garden, and a 'Mary' garden based on plants named after the Virgin Mary. But its most splendiferous element was the thirty-foot-tall scale model of the Matterhorn which rose from the top of the four-acre (!) rock garden, with a summit of alabaster to look like snow, and complete with model cast-iron chamois. This rock garden had an underground cave peopled by more than a hundred garden

*Frank Dobson.

gnomes; indeed, Friar Park was only the second garden in England, after Lamport in Northamptonshire, to play host to gnomes. Crisp was a fine gardener with a great knowledge of alpines, but apparently could not resist a joke: these included a lopsided skull nailed above the scullery door, and friars carved into the terracotta brick of the house.

In 1914, Edward (E.A.) Bowles, who lived and gardened at Myddelton House near Enfield, published *My Garden in Spring*, the first volume of his trilogy. The preface was written by his friend Reginald Farrer, an expert alpine grower and plant hunter but a mischievous fellow. In it, he ridiculed show-off rock gardens: 'This is a mosaic, this is a gambol in purple and gold; but it is not a rock garden, though tin chamois peer never so frequent from its cliffs upon the passer-by, bewildered with such a glare of expensive magnif-icence'. Crisp took tremendous umbrage but, unfortunately, mistook the author and, instead of blaming Farrer, took it out on Bowles. He wrote a pamphlet entitled *Mr E.A. Bowles and his garden, a New Parable of the Pharisee and the Publican*; this was distributed, from a bookmaker's leather bag, at Chelsea Flower Show that year by his friend Miss Ellen Willmott (who in the heyday of her garden at Warley Place employed 104 gardeners). This action caused something of a sensation, and sides were taken. Bowles, who was the most mild-mannered of men, must have been mortified, but restricted himself to a gently reproachful letter to William Robinson, the editor of *Gardening Illustrated*, after the pamphlet was reproduced in it. The row eventually blew over, but it is interesting to note that both Miss Gertrude Jekyll and Professor Henri Correvon of Geneva, the most revered contemporary alpine flower expert, commended Crisp's rock garden, while the garden historian Dr Brent Elliott has called it 'the greatest of all rock gardens'.

At some point after Crisp's death in 1919, Friar Park became a convent and school. My elder brothers and sister attended this educational establishment in the 1950s; my sister remembers eating lunch in a greenhouse, and for years she refused to believe that the Matterhorn was in Switzerland because she knew that it was actually at her school in Oxfordshire. The garden became neglected. The estate was bought by George Harrison in 1971 and has been closed to public gaze ever since. It is said that he has renovated the garden, this monument to Victorian self-confidence and self-consequence, at a cost of millions. In homage to his predecessor he has composed a song, 'The Ballad of Sir Frankie Crisp'. So Crisp has achieved lasting fame after all, if only among ageing pop fans. This may not be quite what he was after; but then, that is the way with fame. I wonder if Greg Dyke would agree.

'A Pleasure and a Privilege'   *14 December 1991*

'I am so lucky to have The Manor. My late husband, Freddie, bought it during the War. It was such a sadness to him that, because of his wretched hay fever, he could not join up, but it did mean that he was able to throw himself into the War effort here at home. And he had the good fortune to pick up the House for a song. Dear Bobo, such a close friend, so old, so gaga, and so grateful to have it taken off her hands, at what was a very difficult time for us all.

'Mercifully, my husband's hay fever never affected his gardening. His War work went so well that he was able to retire in 1945 and devote every hour, when he was not at Newmarket, to supervising the care of his beloved lawns. He used to say, many a time, that he had given his life to the Turf and the turf. Too funny.

'We have been helped since the War by Luigi, you know, who used to be an Italian POW and is now our devoted head gardener. Not a word of English but a wonderful way with *Armillaria*. He is such a dear – but sadly temperamental. We have never been able to bring ourselves to tell him that the War is over, for fear it might quite unhinge him.

'There there is Pam, of course – my dear, down-to-earth sister. She looks after the brassicas and the compost heaps, while I "do" the flower garden. You could not wish for a happier or more fruitful partnership. Such a dear, Pam, so much sadness in her life, yet such a wizard with the aldicarb.

'Though not, at the moment, *quite* at its best, we are frightfully proud of our Rose Garden, where we grow the "old" shrub roses. I would not dream of letting in any of those awful, what I call "Typhoo Tea roses". I told Freddie that I was not having our garden looking like the daily's. So that was that.

'People come here from all over the place to see this garden, which, though I say so myself, is unique. As well as the Rose Garden, we also have a double herbaceous border, a Water Garden and even a Herb Garden! I had this wonderful idea for a White Garden but now *everyone's* got hold of it. I should never have mentioned it to Vita, although she was such a dear, one could not refuse her *anything*. Even "Johnny" Johnston from Hidcote used to drop by for advice. I so remember him asking me how he should trim his hornbeam "stilts". I said, "Johnny, in one word: the pruning knife. Our chap, thingama-jig, swears by them and has only cut himself *really* badly once, I believe." Such a dear, Johnny.

'I rather pride myself on my planting schemes, too. Here is a grouping I particularly like. A wonderful plant, *Alchemilla thingy*. It reminds me of a sort of lady's mantle. If only it were not so difficult to keep alive. I like to be daring and combine it with "catmint" and "lamb's ears". I told Rosemary and Penny about my idea, so now these plants are growing at Barnsley and Tintinhull too! Rather flattering, really. My family say that I am too avant-garde but, as I always say, you have to be true to your creative instincts in this life.

'You will mention we are opening the garden to the public on Bank Holiday Monday, won't you? It is for a good cause:

the GRBS. No, not the Gardener's Royal Benevolent Society, the Guinevere Rakehandle Benefit Syndicate. When is your photographer coming down? Tell him, Mondays are best, when Pam's away running the WI stall at Stow-on-the-Wold market. She is so shy, poor darling; I would not want to put her through it. Tell your man that the best place to take snaps is from behind the delphiniums looking towards the house. On second thoughts, I shall tell him, because that is where I dead-head on Mondays, wearing my old gardening boots. They always come out well in colour.

'I *am* glad you have enjoyed looking round, but I expect you are in a frightful hurry to get home, so will not have time to stay for a cup of coffee or come into the House or anything. You couldn't possibly send six copies of the magazine when the article comes out? Would you? How kind. Better still, ten. Too kind. You must come again . . .'

Guinevere, Lady Rakehandle was in conversation with Ursula Buchan. This article is reproduced by kind permission of the publishers of *Lush Locations*.

Lady Rakehandle's Christmas   *19/26 December 1992*

'I don't mind admitting, I was rather thrilled. Last summer, I was asked by *Lush Locations* whether I would mind if they used my place for their Christmas issue. I agreed, after we had come to an Arrangement, because The Manor does rather lend itself to Festive Occasions. It all came about because, last year, they sent a frightfully tiresome scribbler called Ursula Buchan (sounded like a made-up name to me!) to "write us up". She asked all the wrong questions, went *on* and *on* about the plants, and bored me half to death. I am glad I didn't let her near the House, because she looked just the type who would try to get too chummy.

'However, a perfectly *sweet* photographer came the next week, and he really seemed to understand about this Garden. He hardly trampled on a single delphinium, and took heavenly pictures of our dear little Westie asleep in the trug basket, so it was all right in the end. The magazine people were marvellous: they sent me masses of copies of the article after I rang them up and made a very slight nuisance of myself.

'They had to come and take their pictures for this Christmas thingamajig in August! Some sweet young people arrived and moved all my Things around. They were called "stylists", or some such word, but they didn't look at all like Carly who does my hair in Stow-on-the-Wold on Fridays. You couldn't recognise the dear old place by the end of it.

'Then they brought out wreaths and tartan ribbons and furbelows and cachepots from the car and arranged them all in one corner of the drawing-room, and stuck fir cones on the wall. They even tablescaped the escritoire. They had the dickens of a job finding a Christmas tree in August, as you can imagine, but they finally hit on the plan of sneaking over the ha-ha and digging one up from old Thingummy's plantation, which I must say was jolly enterprising of them. But there was one poor girl whose job it was to spray a fine mist over it all day to stop it dropping needles. Eventually, our daily lent her the spray starch, which did the job beautifully.

'They were all called Sarah, except for a lovely boy called Jacob, with a shaven head and a floppy moustache, who fused all the lights and burst into tears. I think he was a little bit sweet on one of the Sarahs. Too funny.

'The Garden pictures came out the best in the end. It was difficult making the Garden look really wintry but they did a marvellous job with spun sugar and tin cans of snow. Luigi, our darling Italian gardener, was given a waistcoat, a flat cap and moleskin trousers to wear, and spent the entire morning leaning on an antediluvian wheelbarrow which they found buried under the Warfarin in the Fruit Room. He loved it. They had an idea, because of some dam' fool article they had all read in the *Tittle-Tattle*, that these days people like to garden in their birthday suits, but I put a stop to that nonsense right away. I wasn't having him put his back out before the Flower Show.

'I had to send Pam, my dear sad sister, out of the way for the day, because she is inclined to mope about, wearing those ridiculous corduroy breeches and waterproof hat. I told her the compost heaps needed turning, so she went off, as happy as a sandbag, to find her pitchfork. I really didn't think she would find it at all amusing.

'The hardest part was persuading the village school children to exchange their shiny pyjamas and brightly-coloured plimsolls for Sensible clothing, so that they would look right singing carols under the Christmas tree, but a quick telephone call to the Chairman of the Education Committee soon sorted that out. I also had a job digging the Rector out of his pigeon-hide to come and pretend to drink mulled wine. I brought him to heel in the end, though, by hinting I would tell the Bishop what he had said to me about women priests at the Safari Supper in aid of Distressed Commodity Brokers! Altogether, a most satisfactory day, I think.'

Guinevere, Lady Rakehandle was in Conversation with a Very Old and Dear Friend.

8

Contemporary Trends and Preoccupations

Powerplay   *16 March 1985*

Gardening can be tedious, a fact about which gardeners display a reticence which amounts almost to a conspiracy of silence. Many wish to convince the whole world that their hobby is fascinating in all its aspects, but that is not true. It is a liberating feeling to admit that, although one is passionately interested in the cultivation of plants, certain occupations such as weeding, spraying, and mixing compost are not endlessly absorbing. Some gardeners will not acknowledge this even to themselves. Among these I would number the 'power players'.

The typical power player is a middle-aged man, perhaps the senior partner in a firm of solicitors, who lives in a pleasant village a few miles outside a provincial town. He has an acre of ground, including a small orchard and vegetable garden, as productive as his wife and the pensioner who comes once a week can make it. A power player does not need to own a large estate; power-playing is a state of mind.

Gardening bores him, but he likes the idea of it, and finds talking at dinner parties about the water-absorbing properties of polyacrylamides makes a welcome change from the miners' strike. He feels, too, the need to retain some link with the soil, so important, he maintains, to his forebears.

To give credence to his assertion that he is far too busy to weed, he has equipped himself with a leaf-blower, mini-tractor, electric hedge-trimmers and, best of all in his view, a chainsaw. These tools foster in him a sense of importance such as I myself experienced when first let loose with a dumper-truck at Kew. Appreciation of, and pride in, the power and impact of machines overtakes him, and he forgets he is not as strong and fit as he was in those exhilarating days when he stroked Lady Margaret III. On Saturday mornings he calls cheerily to his wife, 'I'm just going out to do a spot of gardening', and leaves to sharpen the cutters on his chainsaw. His wife, though conscious of the attraction of gadgets and prey to a longing for all those useful little tools advertised in genteel mail order catalogues which somehow never quite come up to expectation, sighs slightly, and hunts out the first-aid box. At best she has the prospect of plastering unpleasant little cuts in antiseptic cream, at worst an afternoon spent hanging around in Casualty. Not daring to watch, she hovers just out of sight.

Her husband's chainsaw is an aggressive weapon, subject to spiralling costs like Trident but with no deterrent value. 'Why have it if you do not use it?' he argues. It cuts up, but more often cuts off and, worst of all, cuts down. Every apple tree with the merest hint of canker is severely dealt with, and ominous gaps begin to appear in his orchard.

The chainsaw's saving grace lies in the fact that it is difficult to maintain without proper tools and expertise, and so is often left to rust in a shed after the first occasion when its owner cuts through log, sawing-horse, Wellington boot and cavalry twills.

Anxiety for the welfare of gardeners leads me to counsel caution with machinery – anxiety, and personal knowledge of the harm it can do in the garden. When I was fifteen, the sudden death of my mother brought unwelcome burdens of responsibility onto the shoulders of her children. Domestic help melted away and we were faced with the problem of 'what to do about the garden'. Before that time, gardening for us had consisted of mowing the grass under protest and absently picking daisies out of the lawn while sunbathing and reading novels. As the summer wore on towards autumn we watched, with dismay, the weeds grow alarmingly and threaten to engulf the rose beds and herbaceous borders. My elder brother, with all the confidence and authority of his eighteen years, took a hand. 'What we need', he declared, 'is the flame-thrower.'

Up till then the flame gun had been used gingerly, just on the gravel paths, and the damage had been limited to the scalloping of the adjacent lawn so that its edge resembled a fancy curtain pelmet. Indeed, in those days when weedkiller usually meant sodium chlorate, a material not much less inflammable than paraffin, the flame gun was quite as effective and marginally less dangerous.

How could the jet of fire possibly differentiate, however, between golden rod and sow-thistle? All the vegetation burnt, flaring and dying in much the same way as a firework splutters before finally going out. A wilderness was created in a few minutes and called, provisionally, peace. But worse was to come. Very much in the swing of things now, we pointed the gun at a narrow bed alongside which grew a hedge of *Lonicera nitida*, divided in the middle by a few steps. This hedge was important in that it served to temper the wind to the shorn novel-readers. It might have been dreary, but it was nevertheless venerable and well-covered. The contents of the first border were annihilated satisfactorily, but when we turned the gun a little carelessly on the second half, the flame caught the hedge. There was a sound as of a hundred crisp packets being crumpled, and before anyone could say '*Flammenwerfer*' all that remained were the black, smoking skeletons of several thick trunks with their branches pointing skywards.

Although the fun to be had from garden machinery is quite intoxicating, I freely admit, the power player would be well advised to hire rather than buy it. For if he retains a shred of self-control, and resists the temptation to use it excessively or for the wrong purposes, for months on end he will have to lock his beautiful equipment away in an outbuilding where it will sit depreciating rapidly. If he hires his chainsaw, he knows that before the family can grow anxious about the glint of megalomania in his eye he will be forced to give the damn' thing back.

Round the Mulberry Bush   20 June 1987

Children and gardening go together like peppermint chocolate and red wine: badly. This is odd, because children are natural gardeners; why else should they love sand-pits so? However, even more than other arbitrary and inexplicable adult pursuits, small children get painfully in the way of the business of gardening, proving a danger to themselves and an unwelcome distraction to their bad-tempered progenitors.

It is not, of course, the child's fault that he has yet to learn the ground rules upon which this agonisingly time-consuming activity is based: that most of the time is spent on one's knees pulling out annual nettles (without being stung, by some magic) and jolly buttercups, while leaving alone a mass of drab, stripy leaves; and that, for some reason, only grown-ups are allowed to pick flowers from the borders. The more committed and sophisticated the gardener, the rougher the time had by the child. The blue plastic slide is tidied away without explanation, ball games are kept to a grudging minimum, portions of useful lawn are hijacked and planted up, and flower beds are out of bounds (except to grown-ups, who walk on them all the time). I feel particularly sorry for my own children: instead of idly pulling out the odd sow-thistle while playing Grandmother's Footsteps, I stride out into the garden, armed with notebooks, with a brow not so much furrowed as double-dug by the necessity to do a 'proper job'. In those circumstances, the playful removal or transposition of labels is liable to prompt, in the eyes of the child, a quite disproportionately loud explosion.

It was after a particularly heartfelt outburst that I determined that the gap between what I wished to do and what interested and absorbed the children would have to close. I set aside a piece of ground in the vegetable garden, where my precious colour schemes are unimportant but which is good, well-dug soil and not like the shaded, stony, ill-favoured corner I was given as a child. In it my daughter, aged five, plants fast-growing vegetables like pumpkins and potatoes. The former, provided they germinate, can do with any amount of watering, which is just as well, for watered they will be – about ten times a day. With luck the fruits of her labour will do for Hallowe'en. Potatoes are satisfactory because their progress can be monitored by removing just a little soil to expose the tubers. I bow to tradition and provide seed for her to grow lettuces and radishes (the vegetables most often recommended for children's gardens), but not with much conviction, for children are rarely enthusiastic about growing food they do not eat. Bush cherry tomatoes, which need not have their sideshoots removed, are pretty and useful sown in pots on the window sill. Best of all are the pot marigolds and nasturtiums so resolutely excluded from the garden proper, which germinate like cress, flower without cease, and satisfy the need children have for really bright colours. Nasturtiums are also edible.

Sowing and thinning are, of course, haphazard affairs, but it hardly matters. Much more important, to me at least, is that my children can now recognise common weeds. From the grown-up gardener's point of view, accurate weed identification is vital; once it is learned, the young can be set to a spot of child labour in the big borders. As all they want to do is to feel helpful – for a few minutes, at least – this is hardly too unkind. I allow the use of the best hand tools, although the long-handled Dutch hoe is now banned after an incident involving the greenhouse glass.

Most people with small children will find it no hardship to prune the old common shrubs which form the backbone of many gardens so that safe little hiding places can be made, or to spray the nettles round the bonfire to avoid agonised screams in summer. It is no sacrifice either to plant, in a shady place, the mouse plant (*Arisarum proboscideum*) with its glossy green leaves under which scuttle the long brown 'tails' and humped 'backs' of the curious flowers, the curry plant (*Helichrysum angustifolium*) for the smell of the crushed leaves, the non-flowering form of lamb's ears (*Stachys lanata* 'Silver Carpet') for the furry feel of the leaves, and ornamental grasses, whose seedheads attract the birds. It is scarcely a deprivation to ban the common laburnum with its baleful seeds, or the sinister monkshood, if one cannot depend on one's children not to eat things. There is no reason to be obsessed about all poisonous plants, however, once your children have reached the age of reason; I grew up in a garden full of cuckoo pint, but I cannot remember ever having any desire to eat the orange and red berries.

Interesting children in grown-up concerns like food-gathering and encouraging wildlife should appeal to the incipient adult in them; growing plants for fun appeals to the child in me. If, however, as may well be, my children's interest in their gardening does not last long, I shall buy a few succulent plants, like 'living stones' (*Lithops*) and the spineless sea-urchin cactus (*Astrophytum asterias*), for them to grow in the house; these should take even longer to die of thirst and neglect than the hamster.

Gone to Pot   *2 March 1996*

As a young child my favourite, indeed I should think my only, party-piece was a recitation of that piece of doggerel 'For want of a nail, the shoe was lost; for want of a shoe, the horse was lost . . .' I enjoyed the sense of gathering doom but, even more, I was fascinated (although I may not have expressed myself quite like this at the time) by the way some small incident can detonate a chain reaction of events of greater and greater consequence.

The contemporary vogue for 'container gardening' reminds me of that verse, except that in this case the consequences have been largely salutary, rather than wholly disastrous. What started as an insignificant trend in garden design triggered by an increasing interest in Mediterranean gardens among a few sophisticated gardeners has, in the last few years, grown enough to alter the look of millions of gardens in Britain.

I won't treat you to a paragraph on my feelings about the word 'container'; some word, however ugly, is necessary as a collective noun for tubs, barrels, urns, vases, pots, baskets (hanging or otherwise) and window-boxes, and there is an end to it.

The smallest, but by no means insignificant, consequence is that a variety of previously little-known plants, 'tender perennials', are now widely grown. The tribe of marguerites (*Argyranthemum*) are probably the most popular, but scaevolas, trailing helichrysum, tender convolvulus and many other plants that no one had ever heard of before are now the common coinage of the dimmest garden centre. Nearly all are decorative and positive additions to our gardens. Each year the seedsmen and nurseries offer even more unusual plants which they swear will grow in a basket or a pot. Anything with a droopy habit is a potential winner.

In the past, tender perennials, if grown at all, were propagated each autumn by cuttings, to size them down so that there was room for them under glass during the winter. For many people nowadays this is impossible, but because of the keen interest in container gardening a substantial trade in small rooted cuttings has grown up. These the gardener can buy in the spring, and simply pot up and grow on.

The same is true of tender annuals, or what are loosely called 'bedding plants'. To get the best flowering in containers, you need large, sturdy plants to plant out in them by May. Nurserymen are better at achieving this than most gardeners, especially in the cases of lobelia and pelargonium which need to be germinated 'in heat' in late winter. So fewer and fewer people bother to sow their own seeds of tender annuals, and instead order seedlings and seed-raised small plants, called 'starter plants' or 'plugs', which are delivered in spring. That way they are already past the most dangerous stage of their existence and will happily grow larger on a window sill until the frosts are over.

Both these developments are in accordance with the spirit of the age. If you have long since stopped making your own jam, why should you continue to preserve your own plants, particularly when the nurseries, like the supermarkets, can provide them more cheaply and just when you want them?

The range of varieties which are available as young plants is far narrower than that of seeds in packets. The choice is often limited to the newer, more 'compact' varieties which have been bred to suit containers. This levelling-down is balanced by the fact that garden centres now group for sale plants which will go together happily in a container. 'Recipe' gardening books have promoted this. So, although your choice of container plant may be more limited, you are likely to make a better fist of the colour combinations.

Gardening habits, particularly in propagation, have changed, and there is no reason to think this change temporary. But the biggest impact which pot gardening has had is not on what we buy, but who buys it. A great many people who previously counted themselves out of the gardening lark because they did not possess an inch of workable soil at ground level, or who thought macro-gardening a tiring bore, have taken to growing plants: on balconies, on steps, on paving or on walls. They have joined the gang, and a good thing too.

Would Banks Approve?   16 May 1998

There are many things for which we should be grateful to Sir Joseph Banks, the explorer and botanist. In 1804, together with the son of Josiah Wedgwood, he founded the Horticultural Society of London, the forerunner of the Royal Horticultural Society; he gave his name to *Banksia*, the Australian protea, and *Rosa banksiae*; by accompanying Captain Cook on the *Endeavour* to Australia, he helped to ensure an international audience for Dame Edna Everage; and, rather more controversially, he is credited with inventing the hanging basket.

According to Dr Brent Elliott, the garden historian, he employed them as receptacles for orchids, which he discovered were epiphytic rather than parasitic; that is, they simply clung to trees for support, rather than plugged into them for sustenance. A perforated metal basket, suspended from the roof of a greenhouse or conservatory, must have seemed the best way of mimicking their natural habitat. (To this day, tropical orchids are grown in hanging wire or wooden baskets, filled with tree bark chippings.) However, never in his most exalted fancies could Sir Joseph have guessed how popular they were to become.

All was comparatively well, and quiet, while these baskets hung from roof struts in grand Victorian conservatories, but in recent years they have broken out and made their way into the garden. They now constitute an entire horticultural sub-genre, with a number of plant families associated particularly with them and whole books written about their cultivation and placement. Indeed, the popularity of several tender plants, especially those with a 'trailing habit', depends on their being considered 'hanging basket' plants. Think of *Petunia* 'Surfinia', *Lobelia* 'Cascade', even the tomato with a pendant habit called 'Tumbler'. Not content simply with influencing plant-breeding programmes, hanging baskets have also spawned ill-favoured progeny such as the wall-fixed half-basket, the plastic 'flower tower', and the poly-thene 'flower pouch'.

The enthusiasm with which the British public have embraced hanging baskets in recent years must result from the welcome opportunities they afford people with tiny gardens, or no gardens at all, to grow flowers. But they are far more widely used than that. Instead of being seen as a last resort, they have begun to take centre-stage in many people's summer flower schemes.

There is no doubt that hanging baskets, when well planted with charming summer flowers and so well-watered and fed that the receptacles themselves are hidden, are resplendent objects, capable of catching the eye from a distance and, in dull or dreary circumstances, providing a pleasant 'splash of colour'. However, they almost always lack context, and without that they can look plain silly. The reason I have no objection to planted-up pots on terraces and patios, indeed I positively admire them, is that they are in context, with each other and with the ground from which they almost appear to spring. They soften the hard lines of paving, and act as a link with beds and borders.

Not so with baskets, hanging queasily from their flimsy metal horizontal arms and hooks. They are in relation to nothing, except possibly the next-door basket. They are designed to ornament the house, not the garden; but a good-looking house does not need them, and their very gaiety will draw attention to a plain one, which they cannot hope to hide successfully. Such a house is better off clothed in climbing plants and wall shrubs.

Their use is only really justified where there is no garden soil at all and the urge for colour, life and horticultural fiddling is understandably strong, say in the courtyard of a sheltered housing development or the outside wall of a high-rise flat.

Far be it from me ever to recommend labour-saving for the sake of it, but there is no doubt that baskets can be hard work. They are a fiddle to plant and hang in the first place and, without a pulley system, it is almost impossible not to strain your shoulders, never mind your temper, watering them. And by golly, they need watering. They require regular feeding and deadheading as well. Only by the intelligent use of water reservoirs, water-retaining gel and controlled-release fertilisers can you hope to keep the daily grind, and the cost of metered water, in reasonable bounds.

Despite these strictures, all over the country hanging baskets are at this very moment being planted with half-hardy annuals and perennials for the summer, their owners quite unaware of their inappropriateness. Perhaps I am too harsh? I should love to know what Sir Joseph Banks, a far more open-minded individual than I, thought. It is perfectly possible that the idea of Mediterranean, South African and Far Eastern plants placed in Irish peat compost and English moss and encased in a wire basket from China would have appealed to a globe-circumnavigating naturalist like Banks. You never know.

Good in a Bed   *8 July 1995*

There is nothing more immutable than the mutability of fashion. Even we ploddy old gardeners will pursue anything that we are told is the coming thing. In twenty years, it has been 'All Change' in the flower beds: in place of 'Queen Elizabeth' floribunda roses, golden rod, *Fuchsia* 'Mrs Popple' and *Lobelia* 'Mrs Clibran', we plant 'Graham Thomas' English roses, *Corydalis flexuosa, Astrantia* 'Hadspen Blood' and *Nicotiana* 'Salmon Pink'. So slavish are we in our preference for the voguish that gardens themselves subtly change in appearance.

It is partly the fault of people like me. Always on the look-out for some new angle or possibility, we fail to underline the virtues of the old. When did you last hear anyone suggest you plant a *Lonicera nitida* hedge? In my childhood this strange rela-tion of the honeysuckle, which did not flower and could be close-clipped to make an impenetrable hedge, was as common as privet in southern gardens. It was an invaluable backstop in games of garden rounders, for the branches would catch a ball beautifully. Provided that it is cultivated well, it is still one of the best evergreen garden hedges, yet you would be hard-pressed to find it in any nursery or garden centre.

Those plants which hybridise without trouble, like roses and dahlias, are easy victims of changes in taste. Although much emphasis was placed, after the Second World War, on reinstat-ing the reputation of the (shrubby) Old Roses, long-grown climbing roses have not been treated so well. Pity poor 'Lady Hillingdon', for example. Years ago my brother-in-law, a dis-tinguished politician, discovered that he could get an ice-breaking laugh from a Young Conservative audience with the nursery catalogue description of 'Lady Hillingdon' – 'good in a bed but better up against a wall.' (I should love to know, by the by, with what ingenuity he wove that into a speech on pri-vatisation or the poll tax.) Nowadays, such an audience is more likely to think it a tabloid headline than a description of the good nature of a climbing rose. 'Mrs Oakley Fisher', a once-loved Hybrid Tea rose, has likewise found herself shivering in the cold of public indifference.

As for small trees, I cannot remember when I last saw a young plant of the golden rain tree, *Koelreuteria paniculata*, or the strawberry tree, *Arbutus unedo*; they are both wonderful in their way, but have been widely replaced by *Robinia* 'Frisia' and *Malus* 'Golden Hornet'.

What is strange is how despised are many of those plants which are good enough to receive the Award of Garden Merit; indeed, with the exception of *Lonicera nitida*, all the above-mentioned unpopular plants have this distinction. The AGM is the Royal Horticultural Society's highest accolade for garden suitability and beauty, and it is not handed out with the rations. As the manual puts it: 'The Award of Garden Merit . . . recognises outstanding excellence for garden decoration or use, whether grown in the open or under glass.' How can these plants, tried and tested in many situations and recommended by the RHS's expert Standing Committees, find themselves out of the public eye?

A glance through *The Plant Finder*, which lists most plants carried by British nurseries, is instructive. Those plants which have received the AGM are distinctly marked. You would think that this would be a powerful encouragement for nurseries to stock these plants and for gardeners to buy them. Yet, to take an example at random, the apples 'Belle de Boskoop', 'King Russet' and 'George Neal' are listed in the catalogues of only a tiny handful of specialist nurseries, and unheard-of in the average garden centre.

By such careless regard for what we have, in headlong pursuit of what we are told we should get, do we risk losing part of our plant heritage. Where this has already happened, energetic people must go to great lengths and expense to bring these plants back into general cultivation. Much of the work of the regional groups of the National Council for the Conservation of Plants and Gardens, for example, consists of scouring neglected gardens for once-common plants, and reintroducing and publicising them when they find them. In general, the NCCPG is more interested in plants that are out of cultivation than those which have simply fallen out of favour. However, if we do not take care, the one soon becomes the other.

Mellow Yellow   *20 June 1998*

In glamorous spheres of human activity, like fashion or television, trends spread like wildfire, fanned by the hot, dry winds of ambition and money. Gardening is not like that. Because there is comparatively little money or glamour in gardening, even now, we have the luxury of watching fashions mature slowly, as we might watch a well-made heap of leafmould.

So it has been with the contemporary passion for yellow-leafed plants. Yellow, or yellow-variegated, forms of plain, green-leafed plants have always occurred from time to time, as a result of mutation or 'sporting'. Gradually we have learned to collect and cherish these mutants.

Take a look around your local garden centre. The stand where new plants are displayed will almost certainly contain a substantial proportion of yellow-leafed plants, many of them old friends dressed up in smart new yellow, or yellow-variegated, guise. *Buddleja* 'Santana', for example, is a yellow-variegated sport from the 'butterfly bush', *B. davidii*, while *Spiraea* 'Magic Carpet' is a yellow version of *S. japonica*, and 'Gold Ellen' ("a sister for 'Gold Brian'") a form of the well-known seaside hedging plant, escallonia. You may also come across a new variegated *Ceanothus* called 'Zanzibar' and a yellow-leafed 'bleeding heart' named *Dicentra* 'Goldheart'.

Look further and you will find plenty of others which have

been around a while: a number of yellow-leafed heathers and conifers, together with *Philadelphus coronarius* 'Aureus', *Berberis thunbergii* 'Aurea', *Lonicera nitida* 'Baggesen's Gold', *Choisya ternata* 'Sundance' and, of course, golden privet.

Nurserymen and retailers depend heavily on these plants for adding visual excitement to their displays. So much so, that a traditional mainstay garden shrub like *Spiraea japonica* is rarely sold nowadays, whereas you can buy at least nine very similar gold or golden-variegated sports of it, with alluring names like 'Goldflame', 'Gold Mound', 'Gold Rush' and 'Golden Princess'. Any chance sporting branch, or seedling, is liable these days to be propagated vegetatively, given a name, protected by Plant Breeders' Rights, and then pushed onto the market in a fanfare of publicity. 'Promoted' plants in garden centres are now said to constitute 15 per cent of the trade, so their influence is not negligible.

What is interesting about the current popularity of these plants is that they are demonstrably harder to grow, and to place in the garden. Although the leaves do contain chlorophyll, it is masked by other pigments. This means that a yellow-leafed form of a plant tends to be much less vigorous than its green-leafed counterpart. This can be a positive virtue, especially in the small garden, but it does make these plants harder to rear and maintain, on the whole, and they can be proportionately worse hit by pests. They also revert to green-leafed type all too easily. Those with thin, all-yellow leaves scorch readily in hot sun, yet in the case of, say, *Philadelphus coronarius* 'Aureus', the pretty, white, highly-scented flowers are not nearly so freely borne if the plant is grown in a dark place.

Despite all this, yellow-leafed plants are a very positive element in the garden. They will lift a foliage planting, providing long-lasting and often happy contrasts with green-leafed plants. They can conjure a pool of heartening sunlight in a dark spot, or provide a stimulating contrast to a purple-leafed shrub in a sunny one. It is true that the effects are easily overdone, for too many pools of sunlight undermine the salutary impact of the first, and more than a few stimulating contrasts no longer stimulate but are wearisome and predictable. But knowing that one must use them sparingly does not undermine their essential value.

I wish it were possible simply to lump them all together as desirable garden plants, as the nurserymen do. But, as luck would have it, many of those plants which sport easily also naturally have pink or purple-red flowers, and there is no nastier close-quarters combination than golden-yellow and carmine pink. You have only to think of *Erica* 'Anne Sparkes' or *Spiraea* 'Goldflame'. Just imagine what the new *Buddleja* 'Santana' will look like in August, when it bursts into exciting 'claret' flower. The best yellow-leafed flowering plants, to my mind, are those with strongly blue flowers, like *Caryopteris* ✕ *clandonensis* 'Worcester Gold' and *Veronica prostrata* 'Trehane', which can hold their own against the surrounding yellow leaves – indeed, complement them charmingly.

The real problem, however, lies with their instability. In many yellow-foliaged plants, the cholorophyll fights a brave rearguard action in summer. So among the few things certain in this life (besides death, taxes and Gazza's tears) is the propensity for the leaves of a spiraea to turn from bright butter-yellow in spring to a dirty, dingy lime-green in summer. It is an instructive example of a plant reduced to yallery-greenery.

The Shock of the New   *13 November 1999*

The clocks have gone back, the gardening day has shortened dramatically, and there is finally some time to reflect on those plantings in the garden which are working well, and on those which are not. It may well be the moment to dig out the serial defaulters and the irredeemably second-rate, and think about their replacements.

When faced with the opportunity to replant, I am just as likely to avoid plants I know, as being old hat and without excitement, and plump instead for newly-introduced ones, discovered in the pages of a nursery catalogue, on a bright, cutesy website, or prominently displayed in a garden centre. As if there were not enough plants with which I am already familiar, I find I want those that I have not seen before, for no better reason than that they are unknown. Perhaps you are the same?

Will you, for example, fall for the hype likely to attend the forthcoming launch of a yellow-leafed, red-stemmed climber called *Fallopia baldschuanica* 'Summer Sunrise'? It is supposed to look like the golden hop 'from a distance'. It sounds very interesting, I know; at least, until you recall that *Fallopia baldschuanica* is the current name for *Polygonum baldschuanicum*. Yes, this is a jaundiced version of the Russian vine.

I have shaken my head in the past over the preponderance of yellow-leafed plants among new introductions but, strangely, no one in the nursery trade takes any notice. Purple-flowered buddlejas, pink-flowered spiraeas, blue-flowered ceanothus, all have their yellow or yellow-green-variegated 'sports' which have been pounced upon, named, propagated vegetatively, patented and pushed onto the market in a blaze of publicity. But even the trade must have had moments of doubt about the wisdom of giving a puff to a yellow Russian vine, the cool temperate world's most rampageous scrambler by a mile-a-minute. Although it is not quite as vigorous as the green-leafed version, apparently, I would still lay odds that, if I am foolish enough to plant it, it will be over my fence and clogging up my neighbours' gutters before I can say 'pernicious weed'.

Of course, the trade cannot always be ruled by common sense, for they have a nonsensical public, including me, to please. They know that we expect, and hanker after, a few juicy novelties each year. So convinced are they that wholesale nurseries and seedsmen are prepared to consider giving substantial sums of money to any amateur grower or breeder who has managed to develop a new plant and will allow them to trial and patent it. There is many a plant named after a little old lady whose sharp eyes have picked out a strange foxglove or aberrant geranium in her flower border, and some of the best sweet pea cultivars on the market have been carefully bred by amateurs in their back gardens. Other good plants have been brought home from abroad by accredited plant hunters, trialled and bulked up by specialist nurseries, who then introduce them into general cultivation. Well-known examples of this are the now ubiquitous forms of the blue *Corydalis flexuosa* from China, such as 'Père David' and 'China Blue'.

Even nurserymen would agree that a sizeable proportion of new cultivated plants released onto the market will have a very limited life, sometimes artificially prolonged by publicity and promotions but gradually fading from view because they lack some vital quality which gardeners require. Some have so short a lease that they never enter the consciousness of most gardeners at all. For every *Choisya* 'Aztec Pearl' which succeeds beyond all expectation, there is a *Dianthus* 'Devon Pink Pearl' which comes and goes without remark. Hostas are a good example of a plant group in danger of sinking under the weight of new introductions, some of which are so similar to other named varieties that even the trade is becoming concerned that such a scattergun approach may be self-defeating.

I would not dispute that it is important for plant breeders and introducers to continue to push out the boundaries, for the sake of gene diversity, to accommodate changing social trends, and to help ensure their own commercial viability. After all, from time to time gardeners genuinely benefit. In the last five years a number of highly desirable plants have come on the market, among them the climbing rose 'Penny Lane', *Clematis florida* 'Pistachio', *Phygelius* 'Sensation', some charming and hardy diascias ('Lady Valerie', 'Appleby Apricot' and 'Coral

Belle'), and the French lavenders called 'Kew Red', 'Marsh-wood' and 'Helmsdale'. But we gardeners should use our powers of discernment strictly, and view trade enthusiasms with a sceptical eye. When there is doubt, we should let others test these novelties, if necessary to destruction, and buy them only when they have been widely trialled, in gardens as well as in nursery fields, and are no longer expensive or difficult to find. That is my considered advice, although you could hardly expect me to be guided by it.

Vertically Challenged   30 November 1996

In her novel *Innocence* Penelope Fitzgerald describes a sixteenth-century Florentine family of midgets who surround themselves with equally diminutive attendants in order that the young daughter of the house may be spared the pain of feeling different from the rest of her world. When her midget playmate unexpectedly begins to grow, the daughter decides, from the best of intentions, that her legs should be cut off at the knees. If you read seed catalogues, you could be forgiven for thinking that the plant world has its share of midgets and dwarves, and that plant-breeders are kindly determined to cut most annuals, and even some perennials, off at the knees.

Take the ornamental tobacco plant, nicotiana, as an example. Twenty years ago, forms of this plant available from seed would have been twenty-four to thirty inches tall. Lately, you have been able to buy seed of the 'Domino' and 'Havana' series, which barely reach twelve inches. ('Lime Green', at twenty-four inches, is one of the few old varieties still around in the catalogues.) The newer cultivars have many attributes, such as flowers which stay open in the daytime, but the overall appearance of the dwarves is now unattractively stumpy.

Even more obvious an example is the African marigold, *Tagetes erecta*. The wild plant grows to three feet tall, with a spread of at least twelve inches. Although there are some tall 'open-pollinated' varieties like 'Jubilee' or 'Crackerjack' around, much more publicity is given to the 'F1 hybrids' (which are the result of crossing two selected pure-breeding lines), such as the 'Inca' series, which grow no more than twelve inches tall. African marigolds have such heavy, intrinsically inelegant bobble heads that they need sufficient stem length and a bushy habit to carry them successfully. On a dwarf plant, they look quite out of proportion.

This is where the problem lies. The 'Inca' series of African marigolds are dwarves, not midgets. Although the stem is shortened, the flower head remains the same size, which means a grotesque top-heaviness. In the case of dwarf nicotiana, the leaves have not diminished in the same proportion to the stems.

The taller the original, unrefined plant, the more bizarre it will look when cut down to size. Hollyhocks and sunflowers have been badly stunted. The sunflower 'Sunspot', which has a flower eight to ten inches across, grows on a stem only twenty-four inches tall. You can just imagine how odd that looks. Perennials are not immune, either. Genera which are popular from seed, such as rudbeckias, lupins, dahlias, solidago and coreopsis, have been hammered, or look as if they have.

This has not happened by accident; it is the result of good intentions. In displays of flowers in public places, uniformity and compactness are seen as positive virtues. Compact plants seem suitable for tubs, pots and so on, because they don't need staking, and the flowers cover the plants well. They can be planted densely, which means less frequent weeding and less of a litter problem. Local authorities are bigger customers of seed firms than are private gardeners; therefore, breeding work is aimed more at them than us. It is understandable.

Private gardeners buy 'compact' plants because they are endlessly told by pundits that small plants suit small gardens. Up to a point, Lord Copper. It is a mighty dull garden which has no vertical accents to relieve the patchwork-blanket effect made by massed ground-hugging plants. In fact, even in a small garden it is not necessary to avoid tall plants altogether, only to ensure that most are slender and do not take up too much space laterally.

If, as an amateur, you wish to rebel against this positive discrimination in favour of the vertically-challenged, there are two avenues to explore. You can seek out certain unfashionable genera which have not been 'improved' much: most varieties of cosmos; 'everlasting' flowers like statice and helichrysum; aquilegia; and scabious.

And, if you have time on your hands, there is always the option of collecting the seed of single-flowered, overbred annuals at the end of the season, and sowing it the following spring. Some of the seedlings which appear may well hark back to earlier varieties, perhaps even to the original parents; hang on to the ones you like, throw away the rest, and you could develop your own strain of taller, rangier, better-proportioned plants, more suited to your borders.

Nothing New Under the Sun   27 September 1997

From time to time a phrase or buzzword captures the imagination and attention of a restless, I mean receptive, gardening public. 'Groundcover', 'organic gardening', 'wildlife gardens', 'container gardening' and 'naturalistic planting' have all had their day in the sun. The phrases are a shorthand, imprecise and capable of a hundred interpretations, and they often embody an idea pursued by some people for years. For example, I wrote recently about 'matrix gardening', only to receive a letter from an octogenarian saying that for twenty-seven years she had been trying to save money and effort, and was enchanted to discover that she had, in the process, achieved a matrix garden.

This year, it has been the turn of 'gardening with Mediterranean plants'. Planting a selection of these, it is said, is the right response to the unusual conditions brought about by what is loosely called 'the drought'. On the face of it, there is much to commend this idea, considering that Mediterranean native plants are adapted to endure dry soil and oppressive heat in summer. But we need to define terms. If by 'Mediterranean' we mean the plants that we see in gardens bordering the Mediterranean, then we would be wise not to take much notice.

Like many other Brits, I spent my summer hols in Majorca. In particularly idle moments, when not occupied with the serious business of reading a novel on the beach, I would wander about the resort, looking into villa gardens. They were charming and colourful, but they were also as artificial, in their planting at least, as a seaside garden in Kensington. There were bougainvillaeas from Brazil, *Trachycarpus fortunei* from China, eucalypts from Australia, Canary Island date palms, hydrangeas and *Hibiscus rosa-sinensis* from Asia, cannas from Peru, and plumbago and pelargoniums from South Africa. True, there were also native oleanders and prickly pears, lavenders and cistuses (and, no doubt, plenty of bulbs in the spring), but these were far outnumbered, and outgunned, by flashier families from the subtropics. Even some crops in the fields, such as lemons and figs, were naturalised citizens, not natives.

To take up this point may look like barren pedantry, but I don't think so. I, for one, am much more influenced by what I see than by terse geographical references in plant encyclopaedias. When I think of Mediterranean plants, it is those I have seen on holiday which come immediately to mind. The subtropical exotics in Majorca need a great deal of watering in summer. Every evening, sprinklers and hoses are going full pelt. Where is the advantage in growing these plants if it is shortage of water that worries us? After all, as I was told with glee on my return in late August, while I was away the sun was almost as hot and desiccating in my garden in Northamptonshire as on the beaches of the western Mediterranean.

But the real problem with attempting to grow bougainvillaea, plumbago, Chinese hibiscus, cannas and Canary Island palms is not their thirstiness but the fact that they are not hardy in the British climate. It is the lack of frost in the winter, not of water in the summer, which makes the Mediterranean littoral suitable for these exotics. That, and the wonderfully harsh light which allows the eye to accept their strident colours.

What we should be growing, if we are worried about drought, are xerophytic plants, a number of which are native to the mountains of the Mediterranean region: hairy-leaved sub-shrubs like artemisia, helichrysum, phlomis, cistus, helianthemum and lavender, together with the many bulbs, like species tulips and irises, which solve the problem of how to survive hot, dry summers by dying right down until the autumn rains come.

None of these are strangers to British gardens; that is the irony of it. As with 'organic' or 'container gardening', the idea of gardening with Mediterranean plants is way behind the real times. For many years thoughtful gardeners have been filling sunny borders, where the soil was poor and free-draining, with the hardier 'drought-lovers' from southern Europe, without ever feeling the need to make a great 'do' out of it. The way xerophytes have flourished and flowered in the past few seasons will have been a source of quiet satisfaction to many gardeners not given to grand schemes. And should the very wet and cold winter which is probably just round the corner* succeed in sending such plants to their long home, these people will shrug their shoulders, and wait in quiet amusement to be told the next buzzword. What is it to be? I rather

*In fact, it came in 2000/2001.

favour 'mulching with life'. This phrase has all the virtues: it is easy to misinterpret, of startling modernity (having been invented as late as 1883), and it is practised by people who have never heard the expression.

Round in Circles   14 November 1992

On New Year's Eve, a thousand beacons will be lit across the countries of the European Community to celebrate the advent of the Single European Market. What you may not know – and I think you should – is that you can mark this event in a striking and lasting fashion by planting a tree circle.

The idea is that this circle should be composed of twelve trees – one for every star on the Community flag – as a 'living symbol of European unity'. Notcutts, the tree and shrub nursery who are the official Beacon Europe suppliers of trees to the United Kingdom, are selling plants for three sizes of circle. The small circle must be at least eight metres (that's twenty-five feet in old money, squire) in diameter, the second fourteen metres and the third twenty-five metres. On offer are *Prunus* 'Amanogawa' or *Malus* 'Maypole' for the small circle, *Acer campestre*, *Sorbus aucuparia* 'Asplenifolia', *Malus tschonos-kii* or *Pyrus* 'Chanticleer' for the medium-sized one, and *Fraxinus excelsior*, *Tilia platyphyllos*, *Acer platanoides* or *Prunus avium* for the large one. The idea is that the tree canopies will eventually grow to meet each other, in the most painless kind of 'convergence' likely to occur in Europe.

Anyone wishing to mark this occasion need not restrict themselves to those kinds of tree, of course, although it does make sense to limit the planting to only one species per circle. It would not be a true symbol of European co-operation, after all, if there were to be two-speed circles, especially if some trees grew at the expense of others.

The aim of the tree circle idea is to encourage people, and

local councils in particular, to mark the historic event with a gesture to show their pride in Euro-citizenship. Just picture the excitement generated in the Twinning Sub-Committee of the Amenity and Leisure Committee at Barchester Town Hall when the matter is discussed: a full-dress tree-planting cere- mony in the Councillor Timeserver Memorial People's Park, followed by an all-expenses-paid trip to represent Barchester in an identical event in the Swabian town of Himmel-auf-Erde. It would especially appeal because a tree circle consignment from Notcutts comes complete with a commemorative plaque, thus making the occasion an ideal photo-opportunity for the *Barchester Evening Argus*.

Far be it from me ever to discourage the planting of trees, but a little thought is needed before we commit ourselves, and our gardens, to such an extravagant gesture of solidarity with the European Ideal. It is not that these dozen trees are particularly expensive (including stakes, ties and plaque, they come out at between 247 and 376 écus, depending on the species chosen, exclusive of VAT), but there is an aesthetic difficulty.

I would, frankly, need to be given a basket of narrow-band ERM currencies to persuade me to plant twelve 'Maypole' crab apples in my garden, for these are so-called 'Ballerina' trees, which have no sideshoots, so that they look like nothing so much as leafy poles banged unceremoniously into the ground. *Prunus* 'Amanagowa' is almost the ugliest ornamental cherry on the market. But even those trees on offer which are hand- some, such as the Norway maple, the Callery pear, the broad- leafed lime and the bird cherry, will in almost every ordinary garden, and probably even in parks, look odd if grown in strictly geometric circles. In woodland, the all-important sym- bolic circularity would soon become blurred. Only in a formal setting, where they can be used as a means of defining a very large space, bisected by paths, might they look apposite.

I know we all yearn to give proper expression to our delight at the prospect of European union, but I do not want my descendants, a hundred years hence when my broad-leafed limes are just maturing into imposing trees, to wonder why there is a rondel in the garden. They will, of course, no longer be living in Northamptonshire, but in Danelaw, a backward sub-region of Francogermania. What will they think? That the planting of these trees was part of a pagan ritual, perhaps; or, more likely, marked the passing of the multiples of inches, pennies and eggs. Yes, that must be it. What a sentimental people those 'Britons' were, with their urge to celebrate every transient and unimportant event in their history!

9

Flower Shows

Garden Show   *6 October 1984*

'What's it going to do today?' I asked my fellow judge as we
shook hands, rough palm rasping against rough palm.

'The barograph has been steady since yesterday,' he replied.
'Hot this afternoon, I should think.' He was the head gardener
on a large estate, open to visitors, which ran to rain gauges
and barographs. I felt in my pockets for penknife, soft tape
measure, one-inch-diameter metal ring, and copy of *The
Horticultural Show Handbook 1981*. Our province was the vegetables.

The Handbook, the bible for exhibitors and judges, is not only
technically comprehensive but also adopts a forthright and
high moral tone. For the competitor out for easy glory in a
category where there is little opposition, there is a section
headed 'Prizes not Everything'. Those inclined to blame others
for their own shortcomings it urges, under the title 'Be a

Sportsman': 'An exhibitor who has failed to get a prize and cannot at once see why, should search calmly and patiently for the cause of his competitors' success so that he, himself, may be successful another time.' Should this advice fall on deaf ears, there is a section entitled 'Protests'. It was not our job, however, to award points for the Corinthian spirit.

Inside the grey canvas tent, already hot and airless and filled with the smell of jam tarts and late summer roses, other judges leaned over the central tables, cutting open scones, peering at crochet doilies and sniffing purposefully at bottles of marrow rum. Against the tent walls stood white-clothed trestles of vegetables, fruit, flower arrangements and thin-waisted vases of spiky dahlias, their improbable flowers of flame, carmine and lilac (never orange, pink and mauve) bolt upright on stiff brown stems. I breathed a mental sigh of relief that today they were no concern of ours.

I felt grateful, too, not to be judging the animals sculpted from green peppers and cucumbers, the paintings on dry leaf-skeletons, nor yet 'Flowers in an Unusual Container', in this case a Morphy Richards toaster and a brass flip-top ashtray.

'Cabbages, pointed, two,' I read from the Show Schedule. I plucked up my courage and ventured an opinion. 'Some insect damage?'

'Yes, plenty of meat in this one,' he said, poking about with his pencil inside the outer leaves.

'Whitefly too.' I felt on safe ground, for the life-cycle of *Aleyrodes proletella* was one thing we *had* learned about at Kew.

'Shockingly bad this year; it's the season.'

'A good heart,' I volunteered more boldly.

'Pretty firm.' He pointed rapidly. 'First . . . Second . . . Third. All right by you?'

'All right by me,' I replied, and we moved on.

'Potatoes, round, three.' We held them in the palms of our hands, turning them over to look for signs of scab or the green tinge that indicates too-shallow ridging.

'There's been a visitor here,' he said, pointing to a slug hole. 'These are well-matched. Someone's dug up a whole row.' He picked out three, the colour of day-old cream, with shallow, scattered eyes. 'First . . . Second . . . Third. All right by you?'

'All right by me.' We passed on. 'Three carrots, long-rooted. They're very straight. Did he make holes in the soil with a crowbar?'

'Grew them in a tea-chest, more like; it's roomier. Possibly even used drain pipes. These days it's two-inch diameter plastic pipes for leeks. You can get eighteen inches of blanch that way.' I nodded judicially.

The onions squatted in silver sand, their tops cut, turned down and tied with raffia. The seeds would have been sown in heat in December and the plants set out in raised beds of good rich loam in April. They were larger than a farmhand's fist. We pinched the necks. Some were soft. 'It's the season,' we agreed. Some were shiny, having lost their papery outer covering. 'Excessive skinning,' we said, and marked them down. In the end, sheer size and weight were the deciding factors.

Pickling shallots, displayed in sand on paper plates, were a different matter. That was where the one-inch metal ring came in. Anything larger was disqualified.

'Marrows, table, two.' He was already poking the blade into the end of the first one. A drop of moisture oozed from the cut. He did the same to the others. From one no tell-tale droplet appeared.

'Picked more than a fortnight. Probably done the rounds of the shows.'

Conscientiously we counted the runner beans on each plate. The rules demand twelve, and some competitors cannot count.

The tomatoes, too, were casualties of the season. Many were soft, and some none too fresh, judging by the way the calyces ('spiders') curled in.

Finally, we came to rest in front of the Collections: 'Eight varieties, to be displayed in a box measuring not more than 33 inches by 20 inches.' The tape measures were unwound again. There was no doubt who had won: huge, scrubbed parsnips lay on black velvet, flanked by ruler-straight runner beans and glossy tomatoes; every strand of leek-root had been cleaned and straightened. We were in the presence of a master. But we went through the motions. 'Maximum points, dwarf French beans 15, cabbages 15, leeks 20, lettuces 15 . . .'

'First . . . Second . . . Third. All right by you?'

'All right by me.'

We parted amicably, pleased to have done the task without disagreement, and in the time allowed – for, as usual, the number of entries was down on the previous year.

'Will you be judging at the —— Show next week?' he inquired.

'Of course.' I replied. 'It's the season.'

Glittering Prizes   *23 September 1989*

After more than fifteen years I can still recall the shame and disappointment of being the only girl to leave school without a prize – not even the Low Tackle Shield or, that poisoned chalice, the Effort Cup. Yet last year it was my children who felt shame and disappointment when I lost the Mothers' Race by slowing down, while well in front, for fear of seeming too eager to win.

I decided then that it was time to hone my competitive edge, worn blunt in the intervening years. It is a bit late for me to become a bond dealer, which is why I found myself in early August sitting in the front passenger seat of the car, clutching a decapitated plastic lemonade bottle full of water and sweet pea stems, on the way to enter the 'Four Vases of Three Stems' class at a local show.

All the long, hot, dry summer I had toiled over the sweet pea plants, grown specially for exhibition by the 'cordon' method;

that is, tied individually to eight-foot bamboo canes. My even-
ings were taken up pinching out sideshoots, second leaves
and tendrils, and watering, endlessly watering. It was a bad
summer to grow show sweet peas, and an even worse one to
take the decision not to use insecticides in the garden. The rest
of my spare time was therefore spent squeezing the fat bodies
of greenfly in the forlorn, as it turned out, hope of halting the
spread of debilitating virus.

By mid July I had taken to scanning the plants for sturdy
long shoots holding well-placed flowers, but lack of water at
the roots meant that the stems were pitifully short and, despite
much time spent tying likely candidates loosely to their canes,
they were also rather bent. I consoled myself, as exhibitors
have done since the world began, with the thought that my
rivals would be in much the same boat, although I suspected
that the man who always won buried dead donkeys in his
sweet pea trench in winter. Courteously invited to see his
flowers some days before the show, I almost regretted that my
entry form was already in the post.

On discovering that I possessed no proper green, wasp-
waisted show vases in which to put my flowers, my rival generously lent me four of his and even went to the lengths of
filling them with oasis, leaving two holes into which sweet pea
leaves could be inserted to set off the flowers.

The evening before the show I took a bucket of water up to
the cordons and picked the best of what seemed, to my jaundiced eye, a collection of depressingly short stems, many of
which had three rather than four virus-flecked, sun-scorched
flowers. I put them in a bucket up to their necks and set it in a
cool shed near a window, so that the pollen beetle would
decamp overnight. Some hope.

Arriving at the village hall in early morning, I prepared the
vases outside in the warm sunshine. Although I had brought
plenty of 'spares', there were only just enough good 'blooms'.
They had been cut too late, when they were already fully out,

and it was hard to find ones which were not going past their best. I selected three each of the four varieties and carefully pushed their stems into the oasis, in the shape of a fan: 'Terry Wogan', 'Red Arrow', 'Hunter's Moon' and 'Black Prince'. The vases were placed on the green hessian staging, between 'Floral Art' and gladioli, under the stare of long-dead football teams.

It was three o'clock in the afternoon before the show opened after the judging. Propped up against one of my vases was a blue card – second prize. Later, at the end of the show, I went to give my amiable and victorious friend his vases back. He refused them, saying that if I kept them, he would have some competition again next year. Did he mean 'enough to spur him on, but not quite enough to win'? What he did not realise was that, in the period of waiting, I had discovered that I minded. The edge is back – razor sharp.

This next article is something of a period piece, so much having happened both to parks departments, the care of which is now put out to competitive tender, and to the designs of gardens, thanks in large measure to the influence of television garden makeover programmes. But much is still recognisable.

On Show at Chelsea   8 June 1985

The Great Spring Show of the Royal Horticultural Society, or Chelsea as it is universally known, was once one of the early events of The Season, but the grandeur has gone, since with more than 85,000 Fellows eligible to attend, the Private View is anything but private. The smart will no doubt continue to recognise the smart, but the rest of us can remain cheerfully unaware of a fall in the standard of visitor, if not of exhibit. Chelsea is no longer the place for fine hats and impractical shoes, for spectators must be fit and strong and ruthless to be in any position to enjoy it. There is little pretence anyway that this show is for the visitors, for whom it has become a trial of discomfort and creeping exhaustion.* As a display of gardening technique and achievement, however, Chelsea is beyond all words impressive, especially so after such a very harsh winter and unfriendly spring.

Wood and water were the predominant themes this year. It is marvellous what you can do with some white-painted deal and a small water pump. Everywhere flowering roses climbed over trellis and archway, and wooden bridges spanned a dozen gently flowing rills. Rose nurseries especially could not resist the arbour and the pergola, the bower and the bench. A movement towards what is called the 'natural' but which more properly should be called the 'naturalistic' has gathered pace here in the last few years, echoing what is happening to the collections of plants in botanic gardens, where pot-culture has been widely abandoned in favour of 'landscaped' beds. This mood of 'naturalism', though long present in the rocky outcrops of the alpine nurseries, has brought wild flower meadows and muted harmonious herbaceous gardens to Chelsea, and the wind of change has even ruffled the calm of the National Farmers' Union, so that their stand, once a miracle of cauliflower and carrot pyramids, this year sprouted beehive, lobster-pot and spinning-wheel, and glossy apples tumbling carefully carelessly from a wicker basket.

This naturalism is relative, of course. Chelsea is not fake, far from it, but it is obviously artificial and unrealistic, and no attempt to landscape the exhibits will hide the fact. It stands in the same relation to nature as a reservoir does to a mountain tarn. One's belief can hardly be suspended when snowdrops flower alongside summer lilies; rhodohypoxis from the Drakensberg mountains in Natal share soil pockets with edelweiss from the European Alps; and the City of Birmingham

---

*This was in the days before the RHS restricted numbers of visitors. There are now more than 285,000 Members (once called Fellows) of the Society.

Parks' canal boat drifts between banks of variegated monstera from Mexico and caladiums from tropical South America.

All this could not be otherwise, and hardly matters provided only that no one believes it is possible to copy what is shown to her at Chelsea. This is especially true of the designed gardens outside the main tent, for they depend heavily on high-quality brickwork and paving and elaborate garden structures to which most people simply cannot aspire, and the splashing, tinkling water that flows in such abundance would only encourage the kids to wet their pants or drown themselves.

Late May presents problems for the strict realist in any event, for it is a time of hiatus in our gardens. The tulips are all but over, the roses have not begun. The Great Spring Show, which dates from 1913, must have owed its timing to the influence of the rhododendron growers, so powerful an interest in British horticulture in the early years of the twentieth century. This year, so much have times changed, there was only a handful of exhibits devoted to the rhododendron.

There are old-stagers, of course, who find the new style inappropriate or unattractive, and prominent among these laggards are, not surprisingly, some parks departments (with the heroic exceptions of Belfast and Aberdeen). Half-hardy annuals of ferocious colour intensity still have their place, although large-scale bedding was firmly disavowed by gardening writers as much as a hundred years ago. There was no escaping the rolling masses of schizanthus (the decidedly poor man's orchid) from Slough Corporation, which had all the cloying appeal of a warm Mars bar. Despite the ruinous glasshouse heating bills and labour costs, the popularity of this form of cultivation – Slough won a Gold Medal this year – will ensure its survival.

I enjoy, I must admit, the elaborate conceits of carpet bedding that the Torbay Parks stage every time, for in a show that has everything so that no single one is especially memorable, the thatched cottage of 1983 and the guardsman of last year help me to fix the show in time. This year we were treated to a medieval German house of *Pyrethrum* 'Golden Feather' by the side of the River Weser (blue cineraria), a Pied Piper dressed all in alternanthera, and a distinctly cuddly rat of grey sedum with a mossy tail: a cosy representation of a sinister story, inspired by Torquay's 'twinning' with Hamelin. (What would they have done with Elsinore?)

For me, the most successful exhibits are the small ones: the bulbs, Iceland poppies, alpines, pleiones, African violets, violas and auriculas. They are less wearing and disheartening and more easily comprehended than the large, set-piece stands. They point, too, to a post-war development in gardening. We have become a nation of specialists. Told so often that we have small gardens, we now believe it, and go in far less for a bit of everything.

If this show is primarily for the exhibitors, for whom full order books at Chelsea can mean the difference between profit and loss for the whole year, these growers are nevertheless, at least in the last few days before it begins, caught up in a collective and exultant fellowship that is intoxicating. A respectable pride in excellence sharpened by a healthy commercial spirit results in a spectacle as rich in its entirety as it is exasperating in some particulars, and is a salutary example of the largeness of the human spirit that can contemplate the expending of so much skill, ingenuity and artistry on an event that lasts four short days.

That's Entertainment! *12 June 1993*

There are two reasons why gardeners will almost kill to get to the Chelsea Flower Show each May. The first is that we receive an addictive fix of euphoria when surrounded by so much scent, colour, rarity and exoticism, and this is a perfectly good reason for making the annual effort. The second is that we think we will learn a great deal by being near so much skilful and ingenious design and culture, but this one is based on a false premise.

We are continually told by newspapers and television that we will gain 'ideas to take home and use in [our] own garden'. Garden designers and nurserymen obviously believe it, or they would not take such pains to 'theme' their show gardens and displays. But it is largely a mirage. For unless the theme is single-mindedly laid on with a garden trowel, it will be lost on us. Most visitors to Chelsea are part-time, amateur gardeners, not recent graduates of courses in garden design. How on earth, peering over the heads of 42,000 co-religionists, can they get to grips with some subtleties of interpretation, especially those introduced to enhance the profile of sponsors?

Take this year's Chelsea show as a pretty average example. There were some intriguingly entitled show gardens: 'A Garden of Stories' (*Daily Mirror*), 'A Garden of All Ages' (Action for Blind People), 'Across the Generations' (Help the Aged), 'The Touch of Midas' (Numould and Pershore College), 'The Azzarro 9 Garden' (*Sunday Times*). However, you could not guess that something of the rich history of plant collecting was being shown to you in the 'Garden of Stories', nor yet why Midas's touch required broken balustrading in the Pershore garden. For those of us who have never even heard of Azzarro 9, the fact that 'the feeder fountain spilled on to a diffuser reminiscent of the shape of a perfume bottle' was altogether too subtle.

Moreover, the desire to make a sophisticated design point can lead to errors in planting. There were one or two notable dog's breakfasts, where rhododendron jostled with rose and marguerite smothered astilbe. I hope that these schemes did not end up in too many notebooks.

The themes which were simple and horticulturally apposite worked the best. The National Asthma Campaign's 'Low

Allergen Garden', for example, which was entirely planted with species unlikely to bring on asthma and hay fever, was a good wheeze. Equally successful were the 'dune' garden of wild maritime flowers from Countryside Wildflowers; the 'drought-resistant' garden from Merrist Wood; two 'woodland' gardens, from the *Daily Telegraph* and Bridgemere/WI; the replica of James Pulham's Victorian rock cliff at Waddesdon Manor (*Harpers and Queen*) and *Country Living*'s 'A Celebration of Gertrude Jekyll'. These were sufficiently clear-cut to be genuinely enlightening.

It is not always possible to make something even of the stands in the Marquee. If this was your first Chelsea, you could not know the tricks that were being played on you: for example, that there are no crocuses which will flower at the same time as lilies. The juggling with the seasons which sees snowdrops flower with roses is a bravura example of the nurseryman's skill, and I salute it as such, but it hardly helps anyone to plan a garden.

Of course, there were some straightforward and informative stands. Mattocks succeeded in their attempt to show us what you can do with their 'groundcover' roses (namely, use them in hanging baskets and as standards), Blooms of Bressingham and Notcutts both showed plants for different soils and aspects, and the Alpine Garden Society displayed alpines which grow in woodland. The problem in the Marquee is remembering what it is that you have seen a moment after you have seen it, when carnivorous plants vie for attention with clematis, and Iceland poppies nod at ferns.

Because this show is 'Chelsea', it is easy to be overwhelmed, even misled, by its splendour. Like urchins pressing our noses up against the ballroom window, we are fascinated by the colour, life and accomplishment of it all, but we are not too sure what the toffs are up to. The fact is that the toffs are not always sure, either. Once you accept that Chelsea is not necessarily the place to learn about gardening and garden-making, you can settle happily to the serious and worthwhile business of being entertained by it.

Chelsea Bliss *15 June 1996*

Ever since the days of Lord Aberconway, the last president of the Royal Horticultural Society but two, it has become tradition for each year's Chelsea Flower Show to be described as 'better than ever'. This year it was not. It was nobody's fault. Despite truly Herculean efforts by nurserymen, the very late, cold and dry spring affected the quality of some plants and the timing of their flowering, which meant that carefully planned colour schemes in the show gardens or the stands had sometimes to be abandoned. The only consolation for the visitors was that many flowers in the marquee had the chance to open during the show week for a change instead of, as is more usual, being at their best on Press Day, when only the Queen, her entourage, a couple of thousand corporately-entertained guests and a bunch of hacks get a really good look at them.

What was indubitably better than ever was the arrangement and disposition of celebrities on Press Day. Their numbers have increased markedly at major flower shows, especially Chelsea, in recent years. Their task is often to name newly-bred flowers after a charity, but in some cases it seems simply to add glamour and pizzazz to an occasion which is otherwise just about flowers and gardens and boring things like that. Although their presence probably does not register with really staunch gardeners (who are too busy working in their gardens to be keen watchers of what they probably call the 'goggle box'), they are breakfast, lunch, tea and dinner – well, breakfast anyway – to television producers.

This year there were some intriguing billings, my favourite of which was the promise of Kim Wilde cheek-by-jowl with the Bishop of Norwich. Neil Morrissey and Martin Clunes were there, behaving impeccably, while Betty Boothroyd arrived to name a sweet pea 'Elegance'. Daniel Benzali, 'star of the powerful new courtroom drama, *Murder One'*, was billed to open the BSkyB 'New England Cottage Garden'. A very suitable choice, I thought, for a garden which shouted blueberry-pie American wholesomeness, filled as it was with lilies, lilacs, brick paths, and rocking-chairs on a veranda. If I have a regret about this year's show it is that I did not witness Miss England launching 'Toro Wheel Horse Classic', a charming new rose named after a ride-on lawnmower.

At one moment, a bemused steward asked me if I had any idea who the two 'personalities' disposed photogenically in the National Asthma Campaign's 'Free as Air Low Allergen' garden were. Neither of us could guess, although we could see from the rapt attention of bearded chaps carrying furry muffs on poles that they were not to be sneezed at. (If I were Bill Treacher or Gaby Roslin, I should be mortified to think that I had gone unrecognised by anyone.)

There was a time when Chelsea Flower Show was part of the London Season. Ladies wore hats and gloves and head gardeners accompanied their employers, notebooks in hand. That is a long time ago. Class chic has given way to media chic. If you are the sort of person whose job it is to ask grieving relatives or minor celebrities fatuous questions in a caring voice on daytime television, you can bet your substantial salary that you will one day have to stand in a mock-up garden on the Embankment, directing fatuous questions at tongue-tied garden designers while the wind whips off the Thames and straight up your linen suit.

The BBC has a weakness for choosing presenters for its Chelsea programmes who have never been to the Show before, appear to know nothing about gardening, but are absolutely bowled over by it all. BBC 2's coverage this year included a prime example. She was, of course, loving her first time at Chelsea and, in particular, the gorgeous scent of flowers into which she poked her nose, inhaling deeply: 'Mmmmmm. Bliss.'

I am tempted to say that these things are best left to the experts, in this case the redoubtable and engaging Alan Titchmarsh. But that would be hypocritical of me. For, secretly, I dream of the time when I am a famous or even not-so-famous television celebrity, and I get the chance to appear on screen in a floaty frock, commentating on the Brazilian Grand Prix: 'Brmmmmm. Bliss.'

Overexposure 19 *June 1999*

Garden writers labour in a remote vineyard; it is verdant and fruitful, certainly, but rather off the beaten track. The week of the Chelsea Flower Show, at the end of May, is the only time in the year when we can depend on welcoming workers from elsewhere. From time to time, it is true, magazines and news-papers run news pieces claiming that gardening is the new sex, the new rock and roll, the new 'black', the new carpet bowls, but this concentration on horticulture is spasmodic and temporary. Chelsea is different. The media have always enjoyed Chelsea, for its delicious and predictable cocktail of colour, competitive perfectionism, royalty, and charmingly dotty people to laugh at.

I am very much in favour of extensive coverage of the world's greatest flower show; the more people who are seized with the magic of it the better, and there is always the chance that favourable impressions will linger in the collective con-sciousness, like a catchy song. And the fresh eye often sees what the blasé has missed. However, there is also the danger that tosh will be talked and written. My 'Tosh' award this year went to the journalist who wrote that water featured so exten-sively at Chelsea this year thanks largely to the influence of Charlie Dimmock, who for any space aliens reading this piece is the buxom, wholesome sidekick to Alan Titchmarsh on BBC 1's *Ground Force* makeover programme, a woman who always includes a water 'feature' in the garden plan whether it warrants one or not. Whoever wrote that was not at Chelsea last year, or the year before . . .

There was quite a lot of tosh talked on *Chelsea Live*, the name of a series of eight programmes made by TwoFour Productions and screened by Channel 4, who had exclusive rights for the show. It was anchored by Sue Cook and presented by Monty Don, Diarmuid Gavin, Roy Lancaster and Fiona Lawrenson. That sounds like a strong team of gardeners, but the format consisted of so many short and disparate items that they looked harried from rose pillar to fence post, never able to ask more than a few questions, nor always waiting for the answer. As telly it was amiable, and sometimes illuminating (particularly the more considered inserts of film following people preparing their exhibits elsewhere), but at times it had all the restless inconsequentiality of a butterfly in flight. And an editorial preoccupation with celebrities without conspicuous horticultural credentials resulted in some truly, if probably unintentionally, memorable moments, such as when Adam Faith extolled the virtues of a garden called Great Dickworth, and the Beverley Sisters (Joy, Babs and Teddy), dressed identically, sang 'Anyone else but me', *a capella*.

Too much that was potentially fascinating was not pursued far enough. A discussion about show garden design between Stephen Anderton, gardening correspondent of *The Times*, and the garden designer Mary Keen would surely have been enlightening, had it been allowed to get going, but Sue Cook was far too quick to thank them for coming so she could chase after some other hare. The myriad separate items bumped into each other like the galvanised steel floating plant pots in Paul Cooper's show garden.

The problem was not lack of time, for there were six and a half hours of coverage over ten days, much of it transmitted in the middle of weekday afternoons, filling some of that aching void of post-prandial screen time before *Fifteen-to-One* comes on. (I cannot pretend that I saw more than four hours of it, but that was quite enough to get the flavour.) In fact, too much time was the problem. Chelsea Flower Show simply cannot sustain such prolonged exposure. There was a distinct sound of barrels being scraped. Less would have meant more, and it will be interesting to see whether the RHS thinks so too, when they come to renegotiate the television rights this year.*

Television and other media are important for such an event, not only for showing those who could not go what it was like but for adding to the knowledge and enjoyment for those who did. Which is why it is a pity if the output becomes so cluttered with voguish but tangential preoccupations. I can already see the media coverage for next year's Chelsea Show: 'Due to the enormous success of the film *Star Wars, The Phantom Menace*, designers and horticulturists at Chelsea this year have majored heavily on inter-galactic gardening, with gravity-defying hanging baskets, weightless non-peat potting composts, and "water features" crafted from space-age galvanised steel.' Remember, you read it here first.

The Shows Must Go On   18 July 1992

Harrogate, Ebbw Vale, Malvern, St Helier, Chelsea, Hampton Court, Southport, Edinburgh . . . Defunct marquessates? By-elections won by the Liberals between 1920 and 1955? No, simply some of the major regional or national flower shows and gardening 'festivals' which have been, or will be, held this year.

I sometimes wonder if the much-vaunted British mania for gardening is not more accurately just an enthusiasm for a

*The contract for covering the 2001 Chelsea Flower Show went to the BBC.

specialised spectator sport – a passive version of crown bowls or speedway. Our leisure seems as often spent wandering around the perimeters of huge tents while eating steak sandwiches and talking crossly to our children as in cultivating our gardens. We no longer contribute our own gardening efforts to small local flower shows, and they are dying of inanition as a result. Instead, we go to see what the nurserymen and landscape contractors can do, fleetingly, to entertain and intrigue us.

Horticultural shows may help us while away a few hours but they are fiendish hard work for the nurserymen, especially in a recession when we are all 'just looking'. It is quite possible for them to spend every week in the season setting up a display at a county or regional show somewhere, for all the world as if they were window-dressers, only for a sniffy judge to complain about the spelling on the labels and mark the exhibit down to a bronze. For the majority of specialist nurseries which are two-man bands, that is a hefty commitment, and a potentially disheartening one. To stage a good exhibit means growing, and transporting, at least twice as much as will eventually be needed; after five-day shows, like Chelsea and Hampton Court, many of the plants used will be unfit for sale by the time they are home again.

Of course, no one need exhibit if they do not want to, just as no tabloid newspaper need follow the doings of the royal family, and no Rotarian need turn up at The George on the second Tuesday lunchtime in every month. But choice can seem synonymous with compulsion.

With what mixed feelings, therefore, must nurserymen view the recent increase in 'mega-gardening events'? Last week I went to Hampton Court International Flower Show, only in its third year but already well-established. This is widely spoken of as the summer rival to Chelsea. It is no such thing. It is like a huge, exclusively horticultural county show. Despite the august setting it is informal, friendly, and determinedly all-things-to-all-men. It seemed fitting that, on the day I was there, Radio 2 was broadcasting from the show.

Hampton Court appeals to families, especially those with small children, who are not allowed into Chelsea. But, to its credit, it has managed to avoid fairground attractions; there is no chance of seeing the show through a bilious haze from the top of a Big Dipper. In the end, it is atmosphere more than anything else which separates the two shows. At Hampton Court there is far less of the stuffy, earnest serious-mindedness of the Chelsea crowd, so off-putting to the new visitor, and so comforting to the old.

The nurserymen seem to like it, it must be said, despite its length, because they can sell *plants* there. At Chelsea, only orders can be taken. What is more, although south-west London is hardly free from traffic, it beats Chelsea into a cocked hat for convenience. The show is well-organised, there is masses of space, and the crowds can get to the stands easily.

Unlike at Chelsea, most plants are seen flowering in their due season, although Marks and Spencer could not resist mounting a display of flowering spring bulbs, just to show what you can do with refrigerators and money. This year there were some first-class nursery displays in the six connecting marquees, and it was a pleasure to see so many small nurseries exhibiting a range of summer perennials. However, I found the laid-out gardens, which are such a feature of Chelsea, generally disappointing: with one or two honourable exceptions, they were gimmicky rather than truly imaginative. The requirement that the public be able to walk all round these gardens not only gives them a disembodied look but presents a design challenge which not all can rise to.

Next year we are promised two major new national shows, the RHS Spring Gardening Fair at Wembley at Easter, and the BBC *Gardener's World* show. I trust that the horticultural industry is already thinking up new and exciting ways of entertaining us.

Well-scrubbed Pots and Pans 20 April 1991

If a gap exists between the professional and the amateur gardener, there is a positive mountain chasm dividing the average gardener and the specialist grower of alpine plants. Even the clothes and stance are different: the typical male alpine enthusiast wears a fawn windcheater, patterned jersey and flowing silver beard; both men and women stand with hands behind backs, supporting imaginary rucksacks. Moreover, the crack alpini take their European holidays not in Brittany or the Costa del Sol like the rest of us but in the high, lonely places of the Alps or Pyrenees, where they lie on their bellies photographing androsaces; these pictures are then blown up and displayed for the benefit of their co-religionists at any one of a score of provincial specialist shows. What distinguishes them most from the rest of us, however, is that they elevate attention to detail and care of their diminutive treasures to a high art.

All this was evident at the sixth International Rock Garden Conference, held last week at the University of Warwick. Occurring only once in a decade, this conference is as rare a treat as a glimpse of *Eritrichium nanum* in the Dolomites. It was organised jointly by the Alpine Garden Society (which has local groups in Dublin and Belfast as well as England and Wales) and the Scottish Rock Garden Club. There is something Scottishly self-deprecating about the latter's name, but nothing modest about the size of its membership – 4,500 to the AGS's 13,000 – which shows the particular strength of alpine enthusiasm in Scotland.

It is easy to see why rock plants are much loved in Scotland, particularly away from the wet and humid west coast. Alpines are used to lying snugly under a snow blanket in winter, supplied plentifully with water in the growing season and buffeted by fierce winds all the year round. It is no accident that some of our best rock garden nurseries are in Scotland.

The appeal of alpines can be gauged by the fact that there

were five hundred people happy to spend five days talking about alpines and listening to lectures given by mountaineers every bit as admired by their peers as Bonington or Hillary: Chris Grey-Wilson, Erich Pasche, Jim Archibald, Ron McBeath.

Coinciding with the conference was a competitive flower show. The presence of more than 1,500 well-scrubbed terracotta pots and pans of rock plants made this the largest show of alpines ever held anywhere. The plants' lack of stature did nothing to diminish their impact. Late spring is flowering time for most alpines in our altitudes, and they were *au point*. The colours were strongly though by no means exclusively mauve and yellow, thanks to the many species of primula exhibited, together with diminutive narcissus species and neat cushions of yellow drabas.

The most challenging class of all for the alpine cultivator, the open 'Six pans of rock plants, distinct, no more than two of any one genus', was won by Kath Dryden, the most successful and tireless of exhibitors. She comes from Sawbridgeworth in Hertfordshire, which goes to show that you need not necessarily live in the hills to grow good rock plants.

Few elite alpini have large gardens. Indeed, gardeners often turn to alpines to satisfy their craving for variety when they have too little space for growing much else. It is not a particularly expensive 'hobby', either, for much is grown from seed. Alpine houses, where most of the magical work goes on, are not heated; the point of them is to keep the plants dry rather than warm in winter. Perfection in this field is not, therefore, in theory at least, beyond you and me. If alpines appeal to you but you have never grown them, I would suggest you first join the AGS or SRGC. Both publish, in their journals, invaluable and uncondescending articles for beginners. Next grow a beard.

10

The Natural World

Whither the Weather? — *27 January 1996*

We live in a golden age. Even I, by nature pessimistic, am forced to admit it. Never have so many people gardened with such success using a greater variety of plants, or had more gardens to visit and admire. The amount of information readily available far outweighs any shortfall resulting from the decline in oral tradition. In short, we know a great deal and we have bags of money and leisure to pursue our interest. The only maggots in the apple are the limited size of many gardens, and the cost and unavailability of labour.

Good pessimist that I am, however, I cannot resist pointing out another difficulty: the minute, but significant, shift in the climate. Although annual rainfall has apparently remained remarkably constant in the last three hundred years, winters have lately become perceptibly wetter while summers are drier. Moreover, although temperatures in the growing season have remained the same, winter values have risen by one whole degree Celsius. Which is a lot. (I am indebted for this information to an article by Bill Burroughs, an atmospheric physicist, in *The Garden*, the Journal of the Royal Horticultural Society.)

Of course, most gardeners cannot take such a long view as atmospheric physicists. A relatively stable climate, viewed in the long term, still allows for considerable annual variations, and it is these which influence the way we look at things. They sometimes cause us to jump to groundless conclusions. We view them too readily as trends. Last summer seemed to us then, and was, exceptional, yet there is now a widely held belief that next summer will be similar. On what basis?

If we experience one harsh winter, we fatally lose our confidence in planting half-hardy plants. I know people who won't grow ceanothus any more because they lost them in the hard winter of 1981/82. Ceanothus are not long-lived in any event, and one planted in 1983 would probably have died of natural causes by now. A whole ceanothus lifetime has been lost by timidity.

We must cease relying entirely on our own experience and pay more heed to the measured conclusions of meteorologists. In gardening terms, the fact that winters are tending to be more humid is more important than that summers are drier. We can lay moisture-retentive mulches to help thirsty plants through short-lived summer droughts easily enough anyway; it is more difficult but more necessary to provide a free-draining soil to help nurse, for example, drought-tolerant Mediterranean plants through our wet winters. Changing a soil is harder work than mulching one.

Another consequence of mild, wet winters is that deciduous plants break into bud before it is sensible for them to do so. The chief enemy of promise last season turned out to be that sharp, late frost in mid April, when young growth was already much advanced (the second whammy was delivered by very hot weather following only three months later). Late frosts can make fools even of plants which we normally consider bone-hardy.

There is layer upon layer of meaning in the word 'hardy'. We generally give the title to those plants capable of surviving sub-zero temperatures in winter, but to me that seems too wide a definition. By that token, most hydrangeas are 'hardy'; yet a spring frost can burn their buds and undermine their flowering. Most of us are neither by inclination nor circumstance mollycoddlers of hardy plants. If they are labelled 'tender', we take care to put them in a greenhouse or conservatory. If they are labelled 'hardy', however, we expect them to get on with life without much help from us.

Surely that attitude has to be modified. We must accept the need to protect any slightly dodgy plants in winter, especially those against walls, which will, by virtue of their position, always be more forward. We must also take care with the siting of even hardy plants. East walls which catch the sun first on frosty mornings will shatter the cell structure of many plants which can cheerfully endure any number of cold nights.

We should also look to take fuller advantage of plants from countries with climates similar to the one which we now enjoy: the coasts of New Zealand and Australia, and the eastern seaboard of America, and high altitudes in hotter countries.

Of course, many of the plants readily available in Britain have a wide tolerance of climatic conditions. No plant which becomes universally popular, after all, will do so if it is not universally amenable. So, while enthusiasts like me experiment with hibiscus and agapanthus, the rest of the country happily continues to grow cotoneaster and potentilla, philadelphus and holly, hardy geraniums and aquilegia, secure in the knowledge that neither wet warm winters nor hot dry summers will do for them completely.

For Love of Vanessa   *19 August 1989*

I have a friend who lives at the foot of the Sussex Downs who counts butterflies. For five years, she or her husband (both keen gardeners) has kept a note each day of the species which come to the garden. The day I visited in mid July they counted twelve (Large White, Small White, Green-veined White, Holly Blue, Small Tortoiseshell, Large Skipper, Red Admiral, Brimstone, Meadow Brown, Ringlet, Gatekeeper and Comma), just over half of those they could reasonably expect to see in their garden during the year. It occurred to me that I could not claim such numbers and that perhaps, like many country gardeners, I paid only lip-service to the cultivation of butterflies. Sitting in that garden, with the air constantly stirring from the beat of soundless wings, I felt rather shamefaced.

Which is perfectly ridiculous. Before I get carried away by guilt born of sins of omission, it is well to remember that I, like most gardeners, grow a wide range of butterfly-attracting plants in the garden as a matter of course. Who can honestly say that they have no buddleja? Why, even the brickwork at some London tube stations grows it. The spiraeas, viburnums, lavenders, hebes, valerian, aubrieta, Michaelmas daisies, honeysuckles, ornamental blackberries, sedums and ivies in my garden are all nectar plants.

Geographical and topographical position also influences the range of butterflies in the garden, so concerted encouragement of them will probably increase their numbers more than their variety. The reason I cannot count twelve species on a July day in my garden may be partly because I do not live on the edge of chalk grassland in the south of England.

The real reason for my defensiveness is a common one: a consciousness that encouraging butterflies intelligently demands entomological knowledge of a high order, not to mention an easy familiarity with Jack-in-the-hedge and Yorkshire Fog. But the business can be reduced to manageable proportions, I believe. Plant a wide selection of herbs and allow them to flower; establish sizeable groups of single and preferably throated flowers, many of them pale in colour; leave fallen fruits. Identifying which likes what can come later.

If some of the plants suggested in butterfly books seem a little drab, there are usually more garden-worthy varieties which will do as well: the coloured-foliage forms of the native bugle, *Ajuga reptans*, for example, or *Coronilla glauca* rather than the crown vetch, even June-flowering *Buddleja alternifolia* 'Argentea' and, for a warm wall, *B. fallowiana* 'Alba' as a change from *B. davidii*. I think it is unlikely that the butterflies will mind.

Caterpillars are usually more choosy, but satisfying their wants need not be too complicated. You will need holly and ivy for the Holly Blue; a patch of nettles somewhere inconspicuous for the Peacock, Small Tortoiseshell and Red Admiral; borage for the Painted Lady (not a reference, incidentally, to the little woman and her weakness for Pimm's); spring-maturing cauliflowers because they will survive the depredations of the Large and Small White the previous summer; and a piece of overgrown lawn away from the house for Meadow Brown and Gatekeeper. Most gardens have a flowering currant which their owners cannot quite bring themselves to throw out, despite the smell, which will do for the Comma. The beautiful grass *Deschampsia caespitosa* 'Goldschleier' should please the Ringlet, Speckled Wood and Wall Brown. It has to be said that what these really enjoy is *Agropyron repens*, but not even for the fun of counting butterflies will I purposely cultivate couch grass.

Unsafe Sex   14 July 1990

Scientists working at the Rothamsted Experimental Station have recently isolated the pheromone, or sexual attractant, given off by the female hop aphid. This news gives all too much scope to the British sense of humour. You know the sort of thing: 'Boffins give raunchy greenfly the deadly come-on', or 'Aphids avid for a good time'. As we pick ourselves up off the floor we can only pity the poor scientists, having to isolate the right chemical while simultaneously stuffing their hankies in their mouths.

Joking apart, squire, this work may well have far-reaching consequences for the control of aphids (greenfly and blackfly to you and me), which are some of the worst pests of horticulture and agriculture. They suck the sap from an enormous range of plants and often compound their felony by simultaneously infecting their hosts with virus. The peach-potato aphid, for example, is the vector of Potato Y virus; it is the reason why we have to buy new seed potatoes each year, from Scotland where the aphid does not flourish. What is more, the aphid's powers of reproduction are prodigious. If it were not for all the ladybirds, wasps, hoverflies, lacewings and other predators, we would wade knee-deep in greenfly by the end of an average summer: 'Insect Lotharios eaten alive in population explosion curb bid'.

The work of isolating the particular volatile chemical which is the pheromone from the many which the aphid gives off is a Herculean task; it consists of guiding an incredibly fine tungsten electrode on to the cells (rhinaria) of the male aphid's antennae, which are known to be responsive to stimuli. Chemicals are then blown, singly, over the cells, which react most strongly to the pheromone. To date, the chemical composition of the pheromones is known for the vetch aphid, the black bean aphid, the pea aphid and the wheat aphid, as well as the hop aphid.

In the case of the latter, this was found to be nepetalactol. At the sister experimental station of East Malling in Kent, field trials were carried out using a porous bottle containing volatile nepetalactol suspended above a dish of water. The result was inevitable: 'Death by drowning for disappointed Casanovas'. The scientific methodology for isolating pheromones, though breathtakingly complex, is now well-established, so it should not be too long before we can hang such traps near our rose-bushes and throw away the pirimicarb.

Although the success with aphid attractants is quite recent, for many years a synthetic pheromone which attracts male codling moths, whose caterpillars are the 'maggots' we find in apples, has been available to commercial fruit growers. These days there is even a trap for amateur use, consisting of a corrugated plastic tent with a sticky floor, which can be hung from the branches of a tree. It is said to reduce damage by 90 per cent: 'Randy moths come to a sticky end'. The beauty of pheromone traps is their specificity. Only occasionally, and by chance, will any other insect become trapped in this way.

The only greenfly in the ointment is that aphids tend to reproduce asexually in the summer, only turning to mating in the winter (those long cold lonely nights, eh?) after they have migrated to their primary host plant. Nevertheless, the work is important for the hope it offers of averting a major build-up of the pest, particularly now that some species, such as hop and potato aphids, have become largely immune to insecticides.

Interestingly, pheromone traps are only one aspect of what is loosely called 'biological control', now beginning to become popular with gardeners. Presently available to us is a fungus called *Trichoderma* which is antagonistic to silver leaf disease of plums; a bacterium, *Bacillus thuringiensis*, which can be sprayed on the leaves of brassicas to infect cabbage caterpillars; a mite, *Phytoseiulus persimilis*, which preys on red spider mite in greenhouses; and a parasitic wasp, *Encarsia formosa*, which lays its eggs in the bodies of whitefly larvae, turning them from white to black. If you are having trouble with fly on your toms, I should give 'it' a try.

Nudge, nudge, know what I mean?

Since the article overleaf was written a number of other biological controls have become available to amateur gardeners, such as parasitic nematodes to kill slugs and vine weevils, Cryptolaemus *for mealy bug, and* Aphidoletes *against aphids under glass.*

When the Worm Turns   18 February 1995

I suppose because I am a displaced Celt, it has always seemed to me the greatest blessing to be able to live in Scotland. As a child, having to live permanently in the soft and prosaic south was to me a penance, almost a tragedy, hardly recompensed by yearly holidays in Peeblesshire, the complete repertoire of Andy Stewart records and a fierce devotion to the Jacobite cause. But now I am not quite so sure. At least down here I have not yet had to face the New Zealand flatworm.

For more than twenty years this revoltingly sticky creature has been preying on earthworms in Ireland and Scotland; in some places this has meant the virtual disappearance of the best-known one, *Lumbricius terrestris*, causing problems for farmers, gardeners, birds and small mammals alike. It is only recently, however, that it has breached Hadrian's Wall. There have been at least twenty-five recent sightings in England and more are expected, so southern gardeners have been asked by the Royal Horticultural Society to join their northern and western confrères on Flatworm Patrol.

It is believed that this flatworm was introduced from the South Island of New Zealand, where it occupies a restricted ecological niche, into Northern Ireland, perhaps as eggs in a pot plant, some time at the end of the 1950s. The first positive sighting of it was in 1963, and since then it has spread throughout Ireland and Scotland. It seems to favour gardens, nurseries and grassland in wet, comparatively mild but cool areas. *Artioposthia triangulata* grows up to seven inches long but is usually much smaller; it is triangular in section, purple-brown in colour, buff-coloured underneath, and has a smooth and sticky skin. It moves through the soil in earthworm tunnels or old root channels. It is a nocturnal creature, but can be found curled up under pots or wood over bare soil during the day. The egg capsule is black, shiny, a quarter of the size of a blackcurrant, and contains six to ten young, which are a creamy-white colour when they hatch.

For about ten years it has been known that the presence of this creature in the soil adversely affects the size of the earthworm population and can, therefore, lead in time to poor drainage. I am sure that I do not have to rehearse the virtues of the earthworm. There can be few readers who did not dissect specimens at school, and so discover the beneficial effect it has on the structure of soils and the nutrient distribution therein. Ever since Charles Darwin's study it has stood very high in our esteem, despite its gender-bending proclivities. The idea that a no-good, low-down, foreign . . . well, worm should lie alongside one of these good guys secreting enzymes and sucking up the resulting earthworm 'soup' is absolutely repugnant to us. The truly fiendish thing about this interloper is that it latches solely onto the brightest and best of British soil-dwellers. Why could it not suck up slugs or cutworms, of which we could happily do with fewer?

This worm, which incidentally is no relation to the earthworm, is entirely sinister, having no redeeming features of any kind. This rural terrorist's sole 'purpose of visit' to these shores is to eat earthworms, thereby destabilising agriculture and horticulture and, who knows, perhaps the very fabric of society itself.

It has no natural predators here, and to date there are no chemical control measures approved for use to combat it. If you see one, you are on your own; you will have to stamp on it or put it in a strong salt solution to kill it. The egg capsules can be squashed. You will also have to take guard against its sneaky incursions: washing the compost off container plants, as you probably already do to get rid of that other fiend, the vine weevil, may expose them. Laying polythene or stones on bare earth is a way of trapping them, but it means checking underneath every day. Flatworms are most likely to be spotted in the spring and autumn, when conditions are neither dry or frosty. If past habits are anything to go by, it is likely to be found most in areas of highest rainfall, that is Wales and the west of England, but may occur anywhere.

We are up against a formidable because single-minded and secretive opponent here, but at least, wherever we live, we are all now in this one together. Good luck, and good hunting.

Lush Places   20 July 1991

Our garden is set behind a village street and is of only medium size. Nevertheless, through no virtue of our own, except perhaps the negative one of allowing undergrowth to flourish, it attracts a satisfactory variety of bird visitors to add to the resident population. From time to time we entertain gold-finches, spotted woodpeckers, bullfinches, house martins, swallows and even, one extraordinary October day, a raven, which perched for an hour on the walnut tree before shambling off, Heaven knows where.

We also have more than our share of wood pigeons and col-lared doves, frightened off the cabbage-rich allotments behind the garden by a barrage of plastic supermarket bags. Tied to stakes by orange polypropylene baler twine – that most endur-ing and appealing accessory for the rural man – these billow in a crackling way in the softest breeze.

It would be interesting to discover which brand of plastic bag is the most unsettling to bird life. If I were a bird, I should certainly steer clear of the one given away by Morrison's, for on me it has the same psychological effect as an angry wasp. (Those not fortunate enough to live in the Midlands or North may be unfamiliar with this supermarket chain's black and yellow logo, or indeed with the television advertisement which ends with the memorable jingle 'More reasons to shop at Morreessons'.) It is my unscientific opinion that this bright white bag deters the birds more effectively than the brown and less glaring Sainsbury's bag. Some of these, incidentally, carry the legend 'Age Concern', which indicates a sagacious self-interest on the part of the allotmenteers.

As if the garden did not have enough birds, therefore, the children have pestered me for some time to knock them up a nesting box, going so far as to show me books containing bright, cheerful and simplified illustrations of one in the making. This spring, ashamed of my indolence but lacking the necessary spare bits of wood, I bought a flimsy-looking job for £1.29 at our local 'Pat-a-Pet One-Stop Animal Superstore'. I nailed it, askew, into the bark of the walnut tree.

Nothing happened. All spring the blackbirds, starlings, tits and thrushes studiously ignored it and nested in the 'Albertine' and roof eaves as usual. Then one day, as spring was giving way to a cool windy summer, and long after I had ceased making soothing noises about better luck next year, a pair of spotted flycatchers arrived from North Africa. They inspected the nesting box, saw off some inquisitive starlings, and began to nest there. This was a moment of some excitement, for these birds are not commonplace summer visitors in gardens round here. While the female sat on the eggs, the male would describe a watchful circle round the walnut – kennel, rose arch, trellis, tree stump, kennel. With his flicking tail, white-spotted brown breast and swooping flight, he was a cheering, even an exotic sight.

Last Saturday morning a fluffy fledgling stood on the lip of the nesting box, like a child finding itself for the first time on the edge of the diving board, while a parent fluttered didactically round it. I wanted badly to stay to watch what happened, but even flycatchers could not compete with the West Indies playing at Trent Bridge. The next morning, the garden was strangely dull and still. While we were watching Pringle and de Freitas toil away in the afternoon sunshine trying to bowl out tailenders, the fledgling flycatchers had come through their most dangerous day and were gone. Being no kind of an expert on birds, I had not expected such a swift departure, and my disappointment was keen.

When I look back on this summer, I know that what will stick in my mind is not the unusually dry and cold May, nor yet the continuously wet June, nor even the lateness of the roses which resulted from such an odd passage of weather. For me, this has been the Summer of the Flycatchers. Although I may often choose to ignore the fact, there is, thank goodness, more to gardens than gardening.

A Tale of Tiggywinkles   17 July 1993

This is the time of year when the dead hand of tidiness weighs most heavily on our gardens. Anarchic weed growth is losing its vigour, while the days for uprooting it are long and balmy. Everywhere bare chocolate-sponge soil, punctuated at regular intervals by spindly bedding plants, is encroaching. Shrubs are trimmed, paths are swept, *alles ist in Ordnung*.

Not so here. Untrained climbing rose stems obscure the flowers while shrubs close over narrow paths, their overlying leafage the legacy of close planting and light pruning over many years. Even substantial weeds blush unseen. The garden suffers less from lack of care than from lack of restraint. Infuriating as it must be to orderly people (even I sometimes find the effect self-defeating), it does have the merit of acting as backdrop, cover and larder for wild creatures.

This year I have had several occasions to bless the generous untidiness of my garden. In spring, groundsel left for a week to seed in a rock bed (this, the most successful annual in Creation, will seed at any time of year in open weather) allowed me the pleasure of watching a troupe of five or six goldfinches beating their black and yellow wings within feet of the kitchen door. Conversely, a short-lived rush of tidy-mindedness in April resulted in my scaring away two sitting blackbirds. That kind of thing can spoil my morning as effectively as a tax return.

But the real virtue of a *laissez-faire* attitude was borne in on me when, one fine evening in late June, I came upon four young hedgehogs close to the fire-heap, almost invisible amid the dry rubbish waiting to be burned. Every so often one would shamble off into a large and rather shaming pile of old weeds and garden detritus, left in limbo until I should manage to get round to emptying the two compost bins of their rotted material. This heap was perhaps a reproach to the gardener in me, but it made the ideal nest and feeding ground for hedgehogs.

They must have been less than a month old, their spines were soft and their eyesight appalling. They were not the slightest bit afraid of us: we could come within a few feet before they visibly stiffened. They lumbered round each other, spines touching, like old buffers 'stripping the willow', emitting high-pitched cheeps as they recognised their fellows. One day I found them dozing, snout to snout, in the sun.

I am not keen to promote a welfare dependency culture among hedgehogs. There are far too many slugs on my lettuces for me to want to divert them to catfood. In any event, I am loth to change their natural habits for my own ends. If I want to see them feeding in the evening, I will go looking for them, not force them to come to me. Their appeal for me is their independence, not their malleability.

I feel this especially about hedgehogs because, thanks to Beatrix Potter and the fact that they exhibit characteristics which we like to think are almost human (rolling gait, snuffling nose, bleary eyes), they are prime candidates for anthropomorphism. However, it is precisely because I have resisted this, as being very bad for humans, that I do not consider hedgehogs as spiky but amiable human substitutes and therefore could not regret their leaving, even though the evening trip to the greenhouse thereby lost its savour. One day last week they took off for fresh woods and pastures new, in this case the village allotments, which they could reach by climbing a low wall at the end of the garden and where the slug and insect pickings were probably richer still.

By a mixture of active intervention and benign neglect, it is possible to increase the bird, insect and mammal populations so substantially in the garden as to be noticeable even to the human eye. We can all do it, whether we live in country or in town (hedgehogs, for example, are well known in suburbs and, I strongly suspect, also inhabit neglected ground and allotments in cities), provided that we do not subject our garden to the same level of orderliness which seems appropriate in the house. That is our undisputed domain, but we are only one of many tenants in the garden.

Horse Sense   *2 September 1995*

After 'Machinery', the subject that I disliked most at Kew was 'Turf Culture'. The enthusiasm of the golf course greenkeeper who came in to teach us about bents and fescues, Cumberland sea turf and scarification seemed to me to be extraordinary, even perverse. I saw the point of knowing how to cultivate a good lawn, of course, but the result never seemed to justify the variety and complexity of the tasks involved. Moreover, I was too intoxicated by the enormous possibilities that colourful flowers held out to want to know that making a cricket wicket, caring for a fairway, or managing a horse paddock required the skills and knowledge that I was training to acquire.

Thanks to a fine glowing chestnut New Forest Pony which belongs to my daughter, I have changed my mind. I am now as hooked on 'grassland management' as ever I was on making a flower border. And I cannot deny that what gardening skills I possess have helped to make up for my initial ignorance.

I see now that you cannot, as I had earlier supposed, put a pony in a field and leave it to it. It will eat too much if given the chance and, as it snatches and tears rather than nibbles as sheep do, can uproot the grasses completely in dry conditions. To a trainee grassland manager such as myself, this is both infuriating and worrying. The iron rule about grassland, so I am told by my betters, is that it is either too plentiful and lush, and your hardy native pony becomes fat and prone to laminitis, or too sparse and easily churned up. Horses seem quite incapable of seeing where their best interests lie – which is in preserving their pasture for lean times ahead. Their appeal for people must lie in their blind dependence, because it sure as anything cannot lie in their brains.

Dry summers, such as this one has been, test the patience of horse owners. In our case, we are doubly unfortunate because the paddock is a little small for comfort, and it was only ploughed and sown with a 'ley mixture' last year, having been in a previous incarnation a plantation of Christmas trees. So the grass has not had sufficient time to knit together into a horse-proof mat. We have had to put up an electric tape fence to section off part of the grazing and thereby stop the silly animal eating it all in one huge gorge, and also, once it has been grazed to the height of the square at Trent Bridge, to give it a chance to rest and recover. The paddock is the only grass in our garden, incidentally, which has had expensive metered water sprinkled on it this summer.

My approach to this paddock has been informed by what I know of lawns. I can predict, for example, that when it finally rains I shall witness the same remarkable recuperative power of grasses that is obvious on the lawn. And I can show how useful it is to be able to identify common weeds.

In a garden, ragwort is an occasional nuisance; in a sparse

pony paddock, it is potentially disastrous. Although I am told that ponies will not eat it unless they get very hungry, or it is dried in hay, the consequences to half a ton of horse of its ingestion can be fatal. The thought of the liver damage, photosensitivity leading to severe sunburn, and almost certain death which results from eating ragwort sets me daily scouring the field for seedlings. This is not neurosis, for it is a very common weed and, producing seeds which are carried on the wind, likely to establish itself at any moment. In fact, I have already dug up several seedlings.* The ability to identify plants from their leaves, which gardeners simply take for granted, may possibly one day save the old girl's bacon.

I would not like to give the impression that I resent all this labour. Far from it. For a start, although I do not ride, the pony and I have built up a strong bond since I spend so much time caring for her pasture. And, although this care means precious time taken away from the garden, the favour is not all one way.

One of the aspects of pasture management about which *Horse and Hound* – my guide in all matters equine – is stern is that droppings must be picked up daily, because they can 'taint' the grass in as little as twenty-four hours. So punctiliously every day I wheel them away to rot with old bedding removed from the stable. In future years, I and my neighbours (for there is masses of the stuff) will have the means to improve our rotten heavy alkaline clay garden soil.

What is more, the requirement that horses have for eating what are called 'herbs' is mighty useful to me. 'Herbs', in this case, are not thyme and tarragon but dandelion, yarrow, plantains, clover: in fact, a bunch of garden weeds. Although it

---

*As a result of carelessness about wearing gloves when pulling up ragwort last summer (2000), I suffered a poisoned and very painful thumb for a couple of weeks. I pass this information on so that others will not make the same mistake of treating this baleful weed lightly.

goes against the grain to propagate dandelions, at least I know where I can find some, and it is pleasing to think that the garden's loss is the paddock's gain.

Earthly Immortality   *12 December 1998*

We came upon the place quite by chance. We were staying in the Lake District and had planned to walk up Mellbreak, one of the finest of the Western Fells, which gives its name to the oldest pack of fell foxhounds. As it turned out, the clouds were so low and thick and the rain so insistent that we were forced to stay low, and close to the head of Crummock Water. Which is why we found ourselves walking along a broad path lined on both sides by an elevated ditch and 'hedge' of coppiced ash 'stools' and surrounded by open pasture. These stools were plainly of great antiquity, yet they had the vigour of saplings. Although I am theoretically aware of the astonishing powers of regeneration possessed by plants, I have never seen those powers quite so starkly demonstrated.

Many of these stools were several feet across, covered in moss and hollowed out by decay; yet from a point two feet or so above the ground there rose from the shattered stumps ramrod-straight branches, as thick as a child's arm and in full, generous leaf (this was early October). So empty and cavernous were these stumps that in places it scarcely seemed possible that sufficient continuous threads of phloem and xylem could exist under the bark to feed and water those vigorous ash poles.

I discovered later that these coppiced ash hedges were common once upon a time: carters and travellers would cut down branches as fodder for horses as they passed by, and there was always a demand for ash as firewood. In fact, in the Middle Ages poor men often risked a day in the stocks by stealing hedge wood. Whatever the reasons for this particular avenue's existence, it was a marvellous sight, made more marvellous by the knowledge that coppicing had been carried on here every few years for a very long time, continuing even into our own day.

As Dr Oliver Rackham, the great chronicler of trees and woodland, has written: 'A tree does not have a predetermined lifespan as we do . . . [coppice stools] are completely self-renewing and capable of living indefinitely as long as they are not overshadowed by timber trees.' In other words, they have the potential to be immortal. He goes on: 'An old stool spreads, without loss of vigour, into a ring of living tissue with a hollow centre and often an interrupted circumference.' He cites examples of ash stools on wet sites taking three hundred years to reach two feet in diameter. It was not too fanciful to think, therefore, that these ash had been planted in medieval times.

You may wonder what all this has to do with gardening. That's easy. There are a number of good garden plants which will stand being cut back as hard as those ash, and respond by throwing up new and vigorous shoots. Almost all the native trees and shrubs will do it, of course, but there are some amenable exotics as well, like eucalyptus, catalpa, and cotinus.

The advantage of stooling a potentially large tree such as a eucalyptus by cutting the trunk to within two feet of the ground is that such treatment makes it suitable for growing in a small garden. What is more, eucalypts are shockers for blowing over if allowed to grow to their full height, so are inevitably more stable if grown as stools.

The juvenile leaves of eucalyptus differ from the adult ones (one of the few other examples of this is our native ivy). Coppicing *Eucalyptus gunnii*, and thereby keeping it young, ensures the continuance of the more interesting rounded juvenile leaves, so popular with flower arrangers.

Promoting interesting foliage is the other main reason for coppicing in the garden. The purple- and yellow-leafed hazels (*Corylus maxima* 'Purpurea' and *C. avellana* 'Aurea'), the purple-leafed forms of cotinus (such as *C. coggygria* 'Royal Purple' and 'Grace'), and the coloured-foliage ornamental elders (forms of *Sambucus nigra* and *S. racemosa*) can all be cut to about a foot from the ground in early spring and, if fed and mulched well, will send up strong shoots bearing larger leaves than is usual. This is because coppicing puts roots and crown out of balance, which the plant will naturally try to redress. There is the same amount of root pumping life into a much smaller crown, with the result that the leaves are larger, and may keep their colour for longer in the season.

It is not only a way to promote good foliage; in the case of the shrubby dogwoods (forms of *Cornus alba* and *C. stolonifera*) this spring pruning encourages the growth of young stems, which have the brightest bark colour. You can see these at their best now in 'winter borders' around the country, such as the famous ones at Cambridge Botanic Gardens. I don't suppose I shall look at them again without reflecting that these apparently quite humble life-forms have the theoretical potential for earthly immortality, a fate quite denied to more sophisticated beings like ourselves.

Alien Attack   *19 August 2000*

Two recent and seemingly unrelated news reports have cast a dark shadow on this sunny summer's day. The first, a jokey 'let's have some more fun at the expense of the railways' piece, concerned the particularly severe damage that *Buddleja davidii* has wrought this year to brickwork in stations and embankments. The other news item detailed naturalists' fears that entire ecosystems are now in danger because many native insects are under pressure, or have even become extinct: 7 per cent have lost the battle for survival in the last century.

The buddleja's success at colonisation is partly the result of warm, wet conditions which favour this shrub's growth, and partly because Railtrack can no longer use very strong weed-killers to curb it. The general public is apt to be as sentimental about urban plant survivors as urban foxes, sighing 'Oh, how lovely' when they see purple plumes erupting from the crumbling Victorian walls at London stations. But this shrub creates problems because it can insinuate its roots into any cranny, however small and droughty; moreover, it is an introduced and invasive species, and for that reason alone it should not be tolerated outside the confines of gardens and public parks.

It thrives in the wrong place by default (it was a substantial problem for the railways long before this wet, humid year), as do most of the other aliens which compose two-thirds of our 'native' flora. That's right: there are twice as many introduced plants, mostly garden escapes, in the 'wild' in this country as there are truly native ones.

So what? So an awful lot. Many aliens are benign – think of the beech and the horse chestnut – but there are a number which are serious weeds, notably the aggressively invasive Japanese knotweed, the giant hogweed from the Caucasus which can cause severe allergic reactions in people who touch it, and Himalayan balsam which is just a plain, prolific nuisance, especially along river banks.

The wild plant conservation charity Plantlife and the magazine *Gardening Which?* are presently running a campaign highlighting the dangers posed to the environment by several foreign aquatic plants, sold as ornamental-pond plants in garden centres; these have escaped from gardens and are not-so-gradually clogging up our waterways and replacing native aquatics. The floating pennywort from America (*Hydrocotyle ranunculoides*) develops mats which expand by four inches a day, and is a problem in, for example, the Pevensey Levels in East Sussex; parrot's feather, *Myriophyllum aquaticum*, is a Brazilian plant which grows a floating carpet sturdy enough to support a man, and is quite widely spread now in southern

England, fostered by warm winters; and *Crassula helmsii*, the New Zealand pigmyweed, has colonised five of the nine sites where the native starfruit grows. Tiny pieces of these plants can regenerate and rapidly form new colonies which crowd out better-behaved native plants and deprive fish and other creatures of oxygen. It has been estimated that the annual cost of treating contaminated aquatic habitats is £4–5 million.

As for the disappearance, or drop in numbers, of so many native insects, the reasons cited range from loss of habitat due to building developments or farming practices to poor management of nature reserves. In many instances their disappearance remains a mystery.

However, there is a small unwanted but insistent voice inside me which wonders whether part of this mystery might not be explained by our national devotion to ornamental gardening. There are one million acres of cultivated ground in this country laid out as private gardens, and many are as much a lost habitat for native creatures as a drained peat-bog. Aliens abound, both planted and self-set. With none of these have our native insects 'co-evolved', so they rarely provide sustenance or shelter for more than part of an insect's life-cycle. Buddleja, 'the butterfly bush', for example, is recommended by gardening writers for encouraging butterflies to the garden, as it has tubular flowers rich in nectar in summer, but as a Chinese plant introduced in the nineteenth century it does not suit Peacock butterflies or Small Tortoiseshells as a place to lay their eggs. They choose native stinging nettles instead. The adult Small Comma butterfly may feed on Michaelmas daisies but the caterpillars need dock and sorrel, which we regard as weeds and remove.

There are no doubt a myriad different reasons for the disappearance of insect species, but common sense suggests that we gardeners may be doing our small bit to aid the process: by growing huge numbers of foreign plants, or double-flowered, nectar-poor variants of indigenous species, at the expense of

single-flowered, nectar-rich natives; by failing to create extensive wildlife 'corridors' of locally native plants; and by dump-
ing, wittingly or unwittingly, invasive aquatic plants in the
countryside. Small wonder that, being a gardening writer, I am
prey to uncomfortable thoughts. Well, it will serve me right for
being so smug and thinking I was doing quite a good job . . .

Conversion Tables

For the various Imperial and Metric measurements which appear in the text. These are gardeners' approximations only.

Inches	cm (approx.)
1	2.5
2	5
3	7.5
4	10
5	12.5
6	15
7	17.5
8	20
9	22.5
10	25
12 (1 ft)	30
18	45
20	50.5
24 (2 ft)	60
30	75
33	85
36 (3 ft)	90
90 (7 ft 6 in)	230

Metres	approx. yards	approx. feet
8	8.75	26.25
14	15	46
15	16	50
25	27	82
50	54.5	164
110	120	360

Feet	metres (approx.)
5	1.5
8	2.5
10	3
20	6
30	9
65	20
80	25
100	30

Temperatures °C	°F
20 =	68
10 =	50
0 =	32 (freezing)

Pounds	kilograms
15	6.8
40.5	18.3
74	33.5
710	322
3.5 tons	3.8 tonnes

- 1½ oz. per sq. yd. = 10 gm per sq. m approx.
- 22 yd = 1 chain = 20 m
- ³⁄₁₆ in ≃ ⅛ in ≃ 4 mm

Index

Index

Index

Index